ISOTOPE HYDROLOGY

A Study of the Water Cycle

SERIES ON ENVIRONMENTAL SCIENCE AND MANAGEMENT

Series Editor: Professor J.N.B. Bell
Centre for Environmental Technology, Imperial College

Published

SERIES ON ENVIRONMENTAL SCIENCE AND MANAGEMENT VOL. 6

ISOTOPE HYDROLOGY
A Study of the Water Cycle

Joel R Gat

Weizmann Institute of Science, Israel

Imperial College Press

ICP

Published by

Imperial College Press
57 Shelton Street
Covent Garden
London WC2H 9HE

Distributed by

World Scientific Publishing Co. Pte. Ltd.
5 Toh Tuck Link, Singapore 596224
USA office: 27 Warren Street, Suite 401-402, Hackensack, NJ 07601
UK office: 57 Shelton Street, Covent Garden, London WC2H 9HE

British Library Cataloguing-in-Publication Data
A catalogue record for this book is available from the British Library.

Series on Environmental Science and Management — Vol. 6
ISOTOPE HYDROLOGY
A Study of the Water Cycle

Copyright © 2010 by Imperial College Press

Desk Editor: Tjan Kwang wei

ISBN-13 978-1-86094-035-4
ISBN-10 1-86094-035-8

Typeset by Stallion Press
Email: enquiries@stallionpress.com

Printed in Singapore.

Contents

Chapter 1

The Hydrosphere — An Overview

"Water, water, every where, — nor any drop to drink" (Samuel Taylor Coleridge, "The Rime of the Ancient Mariner") aptly sums up the overall picture of the hydrosphere — that part of planet Earth made up of water. The oceans, covering 71% of the surface of the globe, make up 97.25% of the mass of water. Most of the freshwaters, whose volume is estimated to be $39 \cdot 10^6$ km^3, are also not immediately accessible: $29 \cdot 10^6$ km^3 is ice accumulated on mountain glaciers and on the ice caps of the poles; $9.5 \cdot 10^6$ km^3 constitute groundwaters and only about $0.13 \cdot 10^6$ km^3 are surface waters, mainly lakes and rivers. The amount of water held up in the biosphere is estimated to be $0.6 \cdot 10^3$ km^3. The atmospheric moisture amounts to just $13 \cdot 10^3$ km^3 — less than 10^{-5} of the total amount of water — but this small amount is the one which actuates the hydrologic cycle by virtue of its dynamic nature.

Figure 1.1 shows in a schematic fashion the components of the hydrologic system and the mean annual fluxes between these compartments, i.e. the evaporation, transport through the atmosphere, precipitation over sea and land surfaces, and the backflow to the ocean as surface and sub-surface runoff. Some secondary loops of water recycling from the continents to the atmosphere are also indicated. It is evident that to a first approximation, the hydrologic cycle is a closed one. However, the different reservoirs are not strictly in a steady state, on a variety of time scales. There is a marked seasonal imbalance caused by snow accumulations on large land areas in winter; soil moisture and surface reservoirs such as lakes and wetlands fill up during rainy periods, whereas they drain and dry up or are used up by the vegetation during periods of drought. On a longer time scale, much of the cryosphere and some of the deeper groundwaters are immobilised for long periods and the size of these reservoirs undergo variation on a geological

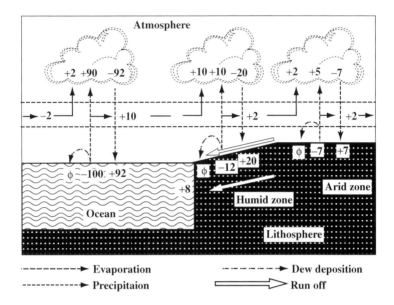

Fig. 1.1. The hydrologic cycle, showing flux units relative to the average marine evaporation rate (100 units). Θ signifies a small fraction of the flux. (Adapted from Chow, 1964).

time scale. In particular, the waxing and waning of glaciers during glacial and inter-glacial periods has resulted in sea-level changes of hundreds of metres.

The total amount of water in the hydrosphere is, however, believed to have been fairly constant throughout most of the geological record, except for the early formative years of the globe. The addition of exhaled water from the interior, by means of volcanism, is nowadays but a very minor factor. Similarly, the loss of water to space, mainly by means of the photolysis of water in the upper atmosphere and the preferred loss of hydrogen atoms, does not amount to much compared to the other fluxes. The residence times or through-flow rates in the various reservoirs are very different, however, ranging from about 10 days in the atmosphere to thousands of years in deep groundwater systems. This concept of the *Residence Time* is further elaborated in Box 1.1.

This text is concerned mainly with *meteoric* waters, i.e. those derived from precipitation, especially those actively taking part in the hydrologic cycle. Thus, the ocean water masses will not be discussed, except as far as they are the sources for the meteoric waters.

Box 1.1 Residence and transit times in water reservoirs.

The residence time of water in a reservoir (τ) is defined as the average time a water molecule will spend in that reservoir. For a well-mixed reservoir at steady state where $F(\text{in}) = F(\text{out})$ so that V=constant [F being the flux and V the volume of the system], this can be expressed by a mass balance equation:

$$\tau = V/F.$$

This time is then equivalent to the one that would be needed to fill up the reservoir. It is further equal to the mean transit of an ideal solute or tracer material, assuming a "piston-flow displacement (PFD)" of the tracer through the medium.

Some average values of the residence time in compartments of the hydrologic cycle are given as follows:

The oceans	3000 yrs. (based on mean evaporation flux)
Groundwaters	500 yrs. (based on the base-flow of the continental discharge)
Rivers	4 months
Atmosphere	10 days

The range of values in each reservoir is very large, especially in groundwater and ice deposits where values can range from a few years in some to thousands of years in others.

The subject has been exhaustively discussed in a number of seminal papers and reviews, more recently in Chapters 9 and 10 in "Solute Modelling in Catchment Systems" (St. T. Trudgill, edtr), Wiley, 1995.

As can be seen in Fig. 1.1, more than 90% of the water evaporated from the oceans falls back as marine precipitation and only about 8% of the evaporated flux is advected onto the land areas. From Fig. 1.2, which shows the distribution of the atmospheric water balance over the globe, we learn that the major source regions of atmospheric moisture are in the subtropical belt. The maximum advective flux of moisture then occurs by eastward flow onto the North-American and European continents at mid-latitudes and by westward transport to the South-American and African continents in the tropical regions. Due to the vertical gradient of temperature in the atmosphere and the resultant low temperatures in the upper troposphere, which limits the amount of water held aloft (Fig. 1.3), most of the vapour is transported in the lower part of the troposphere.

As rain falls on the ground, it is partitioned at (or near) the earth surface into surface runoff and ground infiltration on the one hand, and a return flux of water into the atmosphere by means of direct evaporation or

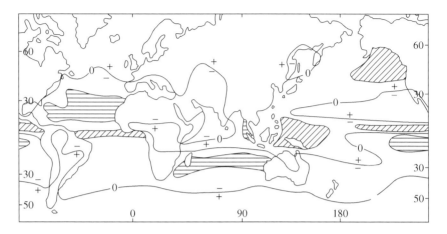

Fig. 1.2. Worldwide ratios of Precipitation/Evaporation amounts. (+) signifies ratios exceeding worldwide average and (−) below that. In stippled areas precipitation excess over evaporation exceeds 100 mm/year and in dashed areas evaporation is in excess of 100 mm/year.

evapo-transpiration through the intermediary of plants, on the other hand. Figure 1.4 schematizes these processes. The major role played by the return flux into the atmosphere is to be noted, which explains the fact that the integrated precipitation amount over the continents exceeds the vapour flux from the oceanic source regions onto the continents. The total amount of re-evaporated waters from all the terrestrial surface reservoirs accounts for more than 50% of the incoming precipitation in most cases and approaches 100% in the arid zone. Details depend on the climate, surface structure and plant cover. The holdup times in the different surface reservoirs prior to evapo-transpiration range from a scale of minutes on the canopy and bare surfaces, to days and weeks in the soil, and up to many years in large lakes.

The potential evaporation, i.e. the maximum rate of evaporation which is that of an open water surface, depends on the climatic condition, the insolation, the wind field and atmospheric humidity. However, since open water bodies occupy just a small fraction of the land surface, it is found that the largest share of the flux into the atmosphere from land is provided by the transpiration of the plant cover, mostly drawing on the waters accumulated in the soil. Evaporation from water intercepted on the canopy of plants also accounts for a surprisingly large share — for example, 35% of the incoming precipitation in the tropical rain forest (Molion, 1987) and 14.2% and 20.3%, respectively, from deciduous and coniferous trees in the

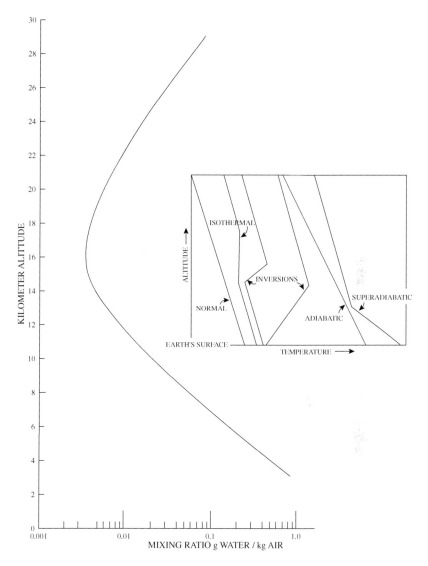

Fig. 1.3. Vertical profile of the mean water content in the troposphere and lower stratosphere. (Inset: typical vertical temperature gradients in the troposphere.)

Appalachian Mountains in Northern America (Kendall, 1993). Direct loss of water by evaporation from the soil, which makes up the balance of the water flux to the atmosphere, is not appreciable where there is an ubiquitous plant cover (Zimmermann *et al.*, 1967).

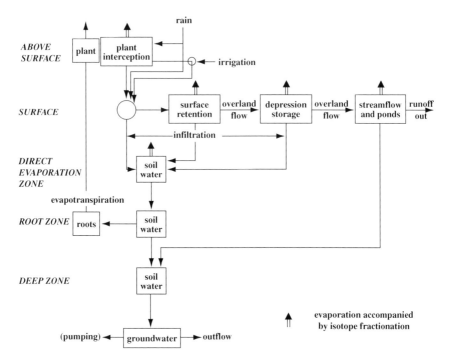

Fig. 1.4.   Scheme of the water fluxes at the atmosphere/land-surface interface (adapted from Gat and Tzur, 1976).

Except for the water recycled into the atmosphere, the precipitation which falls on land ultimately drains back into the oceans, mostly as surface runoff in rivers. However, the travel time from the site of precipitation to the sea is varied, as is the interplay between surface and sub-surface runoff. The latter depends on the climate, land use, morphology and scale of the runoff system.

In the tropics, the major part of runoff takes the form of fast surface runoff, which occurs quite close to the site of precipitation. In the temperate and semi-arid zones, most of the incoming precipitation infiltrates the soil, and that part which is in excess of the water taken up by the plants moves further to recharge groundwaters or to drain to the surface. Most of the groundwater emerges as springs further afield and, as shown in Fig. 1.5, the percentage of surface waters in the total runoff increases on a continental scale. Obviously, some further evaporative water loss can then take place, especially where the surface drainage system is dammed or naturally forms lakes and wetlands. As a rule, the transit times through the sub-surface

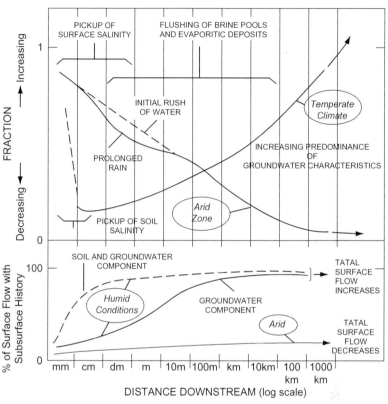

Fig. 1.5. Scheme of the partitioning of the continental runoff between surface and sub-surface flows under different climate scenarios (from Gat, 1980): Top: Fraction of the surface water runoff (corrected for evaporative water losses), scaled downstream from the site of precipitation. Bottom: Percent of the surface runoff with a sub-surface history.

systems are of the order of a few months or years, inversely correlated with the magnitude of the flux.

In contrast, in the arid zone, the largest part of the incoming precipitation is re-evaporated close to the site of precipitation. However, due to the absence of a soil or vegetation cover, even relatively small rain amounts can result in surface flows and, in extreme cases, in flashfloods; these later infiltrate the river bed recharging local desert aquifers. When these waters reappear at the surface they may then dry up completely, forming typical saltpans (locally named salinas or sabkhas). As shown in Fig. 1.5, the

surface to sub-surface relationship as a function of distance from the site
of precipitation in the arid zone differs considerably from that of the more
temperate zones. Moreover, due to the relatively low water fluxes, the ages
of some of the groundwaters of the arid zone are very large, up to the order
of thousands of years.

The quantitative deconvolution of these relationships is one important
task of the tracer hydrology, in general, and of isotope hydrology in
particular.

Chapter 2

The Isotopes of Hydrogen and Oxygen

Hydrogen and oxygen have a number of isotopes, both stable and radioactive. The major isotope of hydrogen with a mass of 1, (^{1}H), occurs in the hydrosphere at a mass abundance of 99.985%. It is accompanied by about 0.015% of the heavy isotope ^{2}H whose particular name is Deuterium, designated D in the older literature. An even heavier isotope of mass 3 (^{3}H), named Tritium, is unstable to β decay with a half-life of 12.43 years; this is the value usually accepted even though more recent measurements suggest the value of $T_{1/2} = 12.23$ years (Lucas and Unterweger, 2000). As this half-life is compatible with the holdup time in many subsurface reservoirs, Tritium is also widely used in hydrologic studies as a tracer.

The most abundant oxygen isotope, ^{16}O, whose mean mass abundance in the hydrosphere is given as 99.762%, is accompanied by a number of stable and radioactive isotopes. The radioactive oxygen isotopes ^{14}O, ^{15}O, ^{19}O and ^{20}O all have half-lives of only seconds and are thus too shortlived to be of any significance in the study of the hydrologic cycle. However, the two stable heavy isotopes of oxygen, ^{17}O and ^{18}O, whose average abundances are 0.0379% and 0.200%, respectively, are the powerhouses of isotope hydrology.

Given the five stable isotopes of hydrogen and oxygen, one will have nine isotopic water molecules (termed isotopologues in the recent literature) ranging from a mass of 18 for ^{1}H$_2$^{16}O to mass 22 for ^{2}H$_2$^{18}O. Assuming statistical distribution of the isotopic species in water, the abundances shown in Table 2.1 are proposed. It is clear that one is concerned mainly with the molecules ^{1}H^{2}H^{16}O, ^{1}H$_2$^{17}O, ^{1}H$_2$^{18}O and, of course, ^{1}H$_2$^{16}O. The doubly labeled molecules ^{1}H^{2}H^{17}O, ^{1}H^{2}H^{18}O and ^{2}H$_2$^{16}O and even more so the triply substituted molecules ^{2}H$_2$^{18}O and ^{2}H$_2$^{17}O can be disregarded at the level of natural abundance of these isotopes.

Table 2.1.   Isotopic water species and their relative abundance.*

Mass	Molecule	Rel.Abundance
18	$^1H_2{}^{16}O$	0.99731
19	$^1H^2H^{16}O$	3.146×10^{-4}
19	$^1H_2{}^{17}O$	$3,789 \times 10^{-4}$
20	$^1H^2H^{17}O$	1.122×10^{-7}
20	$^2H_2{}^{16}O$	2.245×10^{-8}
20	$^1H_2{}^{18}O$	2.000×10^{-3}
21	$^2H_2{}^{17}O$	−negligible
21	$^1H^2H^{18}O$	6.116×10^{-7}
22	$^2H_2{}^{18}O$	−negligible

(* Assuming equilibration during disproportionation reactions)
{Note that the abundance of molecules which are singly substituted by Deuterium,
e.g. $^1H^2HO$, is twice that of the product of the atomic abundances}. As given by
Coplen *et al.* (2002).

Because of the measurement procedure by mass spectrometry, one often
uses isotope (abundance) ratios, R, rather than the conventional concentration units, C. $R = N_i/N_j$ where N_i and N_j are the numbers of the
rare and common isotope species, respectively. For the case of low natural
abundances of the heavy isotope (so that the number of double-labelled
molecules such as ^2H_2O is low) this practically equals the atom-% of the
two isotopes.

The concentration of the isotopic molecules, C, is defined as:

C = (number of isotopic molecules)/(total number of molecules),

and this is related to the isotope ratio as follows: $C = R/(1 + R)$.

One uses a superscript before the symbols to refer to the isotope under
consideration while the molecule involved follows in paranthesis. Thus,
for example, $^{18}R(H_2O)_v$ stands for the ratio of $[H_2{}^{18}O]/[H_2{}^{16}O]$ in water
vapour, while $C(H_2{}^{18}O)_v$, the concentration of $H_2{}^{18}O$, is given by the
ratio of $[H_2{}^{18}O]/\Sigma$ [all isotopic water molecules] which can also be written
as $C(H_2{}^{18}O) = [H_2{}^{18}O]/[H_2O]$, recalling that H or O without any mass
assignment signifies the total amount of that element. In the case of natural
waters where the abundances of the "heavy" isotopes are small, these two
scales do not differ in a significant way.

Even though these isotopes are stable and not subject to radioactive
decay, they can be products or reactants in nuclear reactions initiated
by natural radioactivity or by cosmic radiations. Moreover, hydrogen is
accreted from the solar wind with isotopic abundances quite different from
the terrestrial ones. However, these interactions are believed to be of only

Table 2.2.   Properties of isotopic water molecules.

	$^*\mathrm{H_2}^{18}\mathrm{O}$	$^2\mathrm{H_2}^*\mathrm{O}$	$^1\mathrm{H}^2\mathrm{H}^*\mathrm{O}$
Density [g.ml^{-1} at 30 °C][b]	1.107845	1.10323	1.04945[a]
Temperature at maximum density [°C][b]	4.305	11.24	
Boiling point [°C][c]	100.14	101.42	
Melting point [°C][b]	0.28	3.81	
Self-diffusion coefficient through liquid water [Ð.10^3 (cm^2.s^{-1} at 25°C)][d]	2.66	2.34	
Molecular diffusivity through air relative to the diffusivity of $^*\mathrm{H_2}^*\mathrm{O}$][e]	0.9723		0.9755

Legend:
* — element at natural abundance
a — estimate by interpolation
b — (Steckel and Szapiro, 1963)
c — (Szapiro and Steckel, 1967; Zieborak, 1966)
d — (Wang et al., 1953)
e — (Merlivat, 1978)

minor consequence for the average terrestrial abundance, which can be considered invariant to a first approximation.

The observed variability in the isotopic abundance within the hydrologic cycle, and in other materials in the lithosphere or biosphere, which range about ±50‰ around the mean ^{18}O abundance and from +300‰ to −500‰ in the case of Deuterium (^{2}H), are the result of isotope fractionation processes within the system, to be discussed in the following chapters. Throughout the hydrological cycle, the phase transitions of water between ice, liquid and vapour are the most important processes to be reckoned with.

Some properties of the isotopic molecules of water are shown in Table 2.2.

The ratio of densities of $\mathrm{H_2}^{18}\mathrm{O}/\mathrm{H_2O}^*$ and $^2\mathrm{H_2O}/\mathrm{H_2O}^*$ is close to $d^{30}_{30} = 1.12653$ and 1.108018, respectively. These values are close to the mass ratios of 1.1110 and 1.1117. The good agreement is the result of the fact that inter-atomic distances are almost invariant under isotopic substitution. The molecules $^1\mathrm{H}^2\mathrm{H}^{16}\mathrm{O}$ and $\mathrm{H_2}^{17}\mathrm{O}$ have properties in between those of the heavy and light water molecules.

2.1. Measurement techniques

Mass is the most immediately discernible difference between the stable isotopic species. Thus it is only natural that in the beginning (following the discovery of the stable isotopes of hydrogen and oxygen in the 1930s), the analytical method used for measuring isotope abundance variations in water was based on the density difference between heavy and light water. These measurements were brought to a state of great refinement, so that differences in specific gravity of the order of $0.1\ \gamma$ (one part in 10^7) were detected (Kirschenbaum, 1951). The detection limits by the specific gravity measurements are equivalent to the changes of 0.093 mmol% and 0.083 mmol% for Deuterium and Oxygen-18, respectively. This is well within the range of the natural variability. However, as the density of a given water sample is a function of both the Deuterium and Oxygen-18 content, it was necessary to make allowance for this; for example, by normalising either the oxygen or hydrogen content of the sample.

A number of other analytical methods, which are in use for the analysis of enriched isotopic samples, have also been used at natural abundance levels. Among these methods are the determination of the Deuterium or Oxygen-17 content by n.m.r. and by spectroscopic methods. Indeed the discovery of Oxygen-18 in natural materials by Giauque and Johnston in 1929 was based on the interpretation of the spectrum of atmospheric oxygen, and the first natural abundance determination of Deuterium by H. C. Urey (Urey *et al.*, 1932) was made by measuring the intensity of Balmer lines in the spectrum of hydrogen gas. An additional possible tool is by means of specific nuclear reactions involving one or the other of the heavy isotopes. As an example, the use of the reaction $^{18}O\ (p,n)^{18}F$ was proposed for determining the ^{18}O abundance based on the radiation counting. These methods have the advantage of enabling a (chemically) non-destructive test.

None of all these methods, however, was developed to an extent that would enable the determination of abundances at a sensitivity commensurate with natural variability and, thus, be applicable as routine tools in hydrological studies. Only recently, laser absorption spectroscopy is being developed as an attractive alternative for analysis of isotopic abundances at natural levels (Kerstel and Meijer, 2005).

Mass spectrometry is the most directly applicable tool for isotope analysis. However, in the 1940s, the achievable reproducibility of the measurement of the isotope ratio of $^{18}O/^{16}O$, for example, was barely 10^{-4} and thus not satisfactory for natural abundance variations, which require a resolution of $5\cdot10^{-5}$ or better for observing variations in the oceanic domain. The main reason for the noise in the measurements was long-term instrumental instabilities. This shortcoming was skilfully overcome in the Nier-McKinney mass spectrometer (Nier, 1947; McKinney *et al.*, 1950). This mass spectrometer (MS) was designed specifically for the measurement of small differences in isotope abundance by means of a double inlet (for close comparison between sample and standard) and a double or triple collector so that a direct measurement of the isotope ratio is made. Except for some early attempts at introducing the water directly into the MS (which necessitated a heated and specially treated tube in order to minimize the adsorption and decomposition of the water molecules), one uses more inert gases as carriers of the isotopic signature. The commonly used gas for the measurement of the Deuterium content of the waters is hydrogen gas (H_2), whereas carbon dioxide(CO_2) is preferred for the measurement of oxygen isotopic abundance.

The mass-spectrometric measurement of the hydrogen gas is based on the simultaneous collection of the ion currents for masses 2 and 3. The ion current for mass 2 (I_2) accounts for the ion ($^1H_2$)$^+$; the ion current for mass 3 (I_3) account for the ions ($^1H^2H$)$^+$ and ($^1H_3$)$^+$; to correct for the presence of the ion ($^1H_3$)$^+$ readings are extrapolated to zero pressure in the MS to yield the corrected ion current I_3^*.

It is to be noted that for statistical reasons, the atom ratio of the two hydrogen isotopes, $R(3/2)_{H2}$, is given by the equation $I_3^*/I_2 = 2 \times R(3/2)_{H2}$.

The mass-spectrometric measurement of the CO_2 gas entails the masses 44, 45, 46 and 47.

- The ion current for mass 44 (I_{44}) accounts for the ion ($^{12}C^{16}O_2$)$^+$
- The ion current for mass 45 (I_{45}) accounts for the ions ($^{13}C^{16}O_2$)$^+$ and ($^{12}C^{16}O^{17}O$)$^+$
- The ion current for mass 46 (I_{46}) accounts for the ions ($^{12}C^{16}O^{18}O$)$^+$ and ($^{13}C^{16}O^{17}O$)$^+$
- The ion current for mass 47 (I_{47}) accounts for the ion ($^{13}C^{17}O^{18}O$)$^+$.

The corrected ion current which represents the ^{16}O component (I_{16}^*) is the sum of I_{44} and I_{45} after correction for the contribution of the ion ($^{12}C^{16}O^{17}O$)$^+$ based on the Bigeleisen rule of $\Delta(^{18}O) \approx 2\Delta(^{17}O)$ (Bigeleisen, 1962).

The corrected ion current which represents the heavy oxygen isotope (I_{18}^*) is the sum of I_{46} and I_{47}. Usually, I_{47} is not measured and its amount is estimated according to the contribution of the ^{13}C atom based on the ratio of I_{45}/I_{46}. The atom ratio of the isotopes $^{18}O/^{16}O$ is then given by the equation:

$$I_{18}^*/I_{16}^* = 2 \times R(18/16)_{CO_2}.$$

The procedures for preparing the samples of water for analyses (i.e. the conversion to CO_2 and H_2 respectively) is described in Box 2.1. Some special features of analysis of saline waters are described in Box 2.2. Lately, the use of carbon monoxide (CO) and oxygen gas is becoming more prevalent, using online conversion techniques. The measurement of the rarer isotope ^{17}O will be described in Box 2.4.

Box 2.1. Procedure for preparing water samples for MS analysis.

2.1a. Procedures for analysis of the hydrogen isotope abundance:

(1) Classically, water vapour is obtained by the distillation of water from the sample, followed by decomposition of the vapour over hot metal, e.g. Uranium, Cobalt or Zinc. Assuming no fractionation during distillation, then $^2R(H_2) \sim {}^2R(H_2O)$. For problems encountered in the case of saline waters, consult Box 2.2.

(2) By equilibration with hydrogen gas over a catalyst at a well-defined temperature, to attain equilibrium according to the reaction:

$$^1H_2(g) + {}^1H^2HO \text{ (liq)} = {}^1H^2H \text{ (g)} + {}^1H_2O \text{ (liq)}.$$

In this case, $^2R(H_2) = {}^2\alpha^*(\text{T.Sal}).{}^2R(H_2O)$ after correction for the Deuterium introduced by the hydrogen gas, where $^2\alpha^*$ is the equilibrium fractionation factor for the Hydrogen-Water exchange reaction. Note that α^* is a function of both temperature and the salinity.

Note that the isotope abundance will be given in the concentration scale in the case of preparation 1 and in the activity scale for case no. 2.

2.1b. Procedures for measuring the $^{18}O/^{16}O$ ratio in water by conversion to CO_2 gas:

(1) Usually by equilibration with CO_2 (acid catalysed) according to the equation:

$$H_2{}^{18}O \text{ (liq)} + C^{16}O_2 \text{ (g)} = H_2{}^{16}O \text{ (liq)} + C^{16}O^{18}O \text{ (g)}$$

After correcting for the isotopes introduced by the CO_2 gas, we have:

$$^{18}R(CO_2) = {}^{18}\alpha^*(\text{T,Sal}) \cdot {}^{18}R(H_2O)$$

where $^{18}\alpha^*$ is the equilibrium fractionation factor for the CO_2–Water exchange;

(2) Decomposition and synthesis by the following sequence of reactions:

$$H_2O \rightarrow (Br_xF_y) \rightarrow O_2 + HF; \quad O_2 + C \rightarrow CO_2.$$

In this case, $^{18}R(CO_2) \approx {}^{18}R(H_2O)$ on the concentration scale.

Box 2.2. The isotope salt effects and the analysis of saline waters.

The fundamental feature of stable oxygen and hydrogen isotopes ($^{18}O/^{16}O$ and $^{2}H/^{1}H$) in saline waters is that the thermodynamic activity ratios of these isotopes differ from their composition ratios, which is known as the "isotope salt effects." The isotope salt effect, first described by Taube (1954), results from the interactions between dissolved electrolyte ions and water molecules (e.g. hydration of ions) which change the activity of isotopic water molecules. The isotope salt effect can be rigorously defined in terms of thermodynamics (Horita *et al.*, 1993), as in Eq. (1):

$$\Gamma = \frac{R_{activity}}{R_{composition}} = \frac{a(HDO)/a(H_2O)}{X(HDO)/X(H_2O)}$$

or

$$\frac{a(H_2{}^{18}O)/a(H_2{}^{16}O)}{X(H_2{}^{18}O)/X(H_2{}^{16}O)} = \frac{\gamma(HDO)}{\gamma(H_2O)} \quad \text{or} \quad \frac{\gamma(H_2{}^{18}O)}{\gamma(H_2{}^{16}O)} \tag{1}$$

where a, X, and γ denote the activity, mole fraction, and activity coefficient of isotopic water molecules, respectively. The R stands for $^{18}O/^{16}O$ or $^{2}H/^{1}H$. From the relation $R = 1 + 10^{-3}\delta$, we obtain,

$$10^3 \ln \Gamma \cong \delta_{activity} - \delta_{composition} \tag{2}$$

At any given temperature, the magnitude of the isotope salt effects is practically linear with the molality (mol/kg H_2O), and the oxygen and hydrogen isotope salt effects in chloride-type brines at 20–25°C can be expressed as a sum of the effects by individual electrolyte components (Sofer and Gat, 1972, 1975; Horita, 1989), as in Eq. (3):

$$10^3 \ln \Gamma(D/H) = 2.2m\,NaCl + 2.5m\,KCl + 5.1m\,MgCl_2 + 6.1m\,CaCl_2$$
$$10^3 \ln \Gamma(^{18}O/^{16}O) = 0.16m\,KCl - 1.11m\,MgCl_2 - 0.47m\,CaCl_2 \tag{3}$$

where m is the molality (mol/kg H_2O). Horita *et al.* (1993, 1995) extended experimental determinations of the isotope salt effects to elevated temperatures.

For isotopic fractionation processes between brines and other phases (vapor, gas, minerals) and dissolved species, the isotope activity ratio should be used. For processes in which isotopic water molecules behave conservatively (e.g. mixing of different water bodies, dissolution-precipitation of evaporite deposits), the isotope composition ratio is the unit of choice. However, even in such cases the isotope activity ratio is equally useful, because the magnitude of the isotope salt effects is practically linear with the concentration of dissolved salts. For more in-depth discussion and examples for geochemical implications of the isotope salt effects, see Horita *et al.* (1993b).

The isotopic compositions of saline waters have traditionally been determined by the same methods which have been used for fresh waters; metal- (Zn, U, Cr...) reduction for $\delta^2 H$ and CO_2–water equilibration methods for $\delta^{18}O$ values.

(*Continued*)

Box 2.2. (*Continued*)

It should be noted that the former method yields the $\delta^2 H$ composition values, but that the latter method the $\delta^{18}O$ activity ratio. In addition to this discrepancy between the two analytical methods, several problems are encountered in the isotopic analysis of brines due to dissolved salts, e.g. incomplete reduction of water in brines to H_2, sluggish isotope equilibration between CO_2 and brines, etc. (Horita, 1989). The former problem, particularly, resulted in poor $\delta^2 H$ values of hypersaline brines from the Dead Sea. Using the H_2-water equilibration method for obtaining the $\delta^2 H$ activity ratio (Horita, 1988) and the azeotropic distillation method after the removal of alkaline-metal cations for obtaining the composition ratio (Horita and Gat, 1988), it was possible to determine directly both $\delta^2 H$ values for a suite of Dead Sea brines (Horita and Gat, 1989). These results show that the $\delta^2 H$ composition values of Dead Sea brine from the literature are systematically (up to ca. 10‰) lower than the new set of data. In addition, the new data of $\delta^2 H$ varies linearly with $\delta^{18}O$ activity values as expected for evaporative water bodies. The directly measured $^2 H/^1 H$ isotope salt effects ($10^3/\ln\Gamma$) for the Dead Sea brines of salinity $\sigma_{25} = 232.2$ (mNaCl=1.95, mKCl = 0.15, mMgCl$_2$ = 2.00, and mCaCl$_2$ = 0.49) is +17.8‰, which is in excellent agreement with that calculated from the experimentally determined Eq. (3). A small (+1.2‰) hydrogen isotope salt effect was detected even for a suite of seawaters (Shank *et al.*, 1995).

The H_2 — water equilibration method is a preferred method for $\delta^2 H$ measurements of brines because of (a) high precision and accuracy, (b) yielding the isotope activity ratio, and (c) high throughputs by automation (Horita *et al.*, 1989).

Using the differential MS measurements, reproducibilities of better than $2 \cdot 10^{-5}$ for the $^{18}O/^{16}O$ and $^{13}C/^{12}C$ ratios were achieved. The preparation of the sample for mass spectrometric analysis, rather than the MS measurement itself, is often the weakest link in the analysis.

The differential measurement technique obviously requires a reference, with which the measurements of the different samples are compared. Following some early arbitrarily chosen standards, such as the NBS–1 reference standard (a sample of Potomac River water), Craig in 1961 suggested the mean ocean water composition as a natural standard for the measurement of the water isotopes, in view of the pre-dominance of ocean water in the hydrosphere. However, even though the differences in isotopic composition among the various oceanic water masses are relatively small (of the order of 1‰ in ^{18}O abundance), a virtual "standard mean ocean water" (SMOW) was defined relative to the NBS–1 standard as follows

(Craig, 1961):

$$R\left(^{18}O/^{16}O\right)_{SMOW} \equiv 1.008.R\left(^{18}O/^{16}O\right)_{NBS\text{-}1};$$

$$R\left(^{2}H/^{1}H\right)_{SMOW} \equiv 1.050.R\left(^{2}H/^{1}H\right)_{NBS\text{-}1}.$$

Later, an actual water sample which can be measured and whose isotopic composition matches that of the defined SMOW was made up and deposited at the International Atomic Energy Agency (IAEA) in Vienna. Nowadays, all measurements of the isotope abundance in water are reported in comparison to this VSMOW reference standard. In view of the relatively large range of the isotopic compositions in the hydrologic cycle, which often exceeds the linearity of the mass spectrometers, a number of secondary standards have been prepared for operational reasons and their assigned values are shown in Table 2.3. In this table are also shown standards for these isotopic species and of other elements in substances other than in water itself.

Due to this measurement procedure, the isotopic abundances are reported, in practice, as δ values, i.e. the relative deviations with respect to the standard value, as defined as below:

$$\delta = \left(R_{sample} - R_{standard}\right)/R_{standard} = \left(R_{sample}/R_{standard}\right) - 1.$$

Applied to the deuterium–hydrogen pair, the notations of either $^{2}\delta$ or $\delta(^{2}H)$ is used. The equivalent notation for the pair of isotopes of ^{18}O and ^{16}O is $^{18}\delta$ or $\delta(^{18}O)$. In older literature, ^{2}H was referred to as D (short for Deuterium) and the notation δD was used.

Table 2.3. **Standards used for calibrating the δ-measurements.**

Name of Standard	Isotope/substance	assigned δ-value ‰
VSMOW	^{18}O in water	$^{18}\delta = 0$ [vs SMOW]
VSMOW	^{2}H in water	$^{2}\delta = 0$ [vs SMOW]
GISP	^{18}O in water	$^{18}\delta = -24.79$ [vs SMOW]
GISP	^{2}H in water	$^{2}\delta = -189.7$ [vs SMOW]
SLAP	^{18}O in water	$^{18}\delta = -55.50$ [vs SMOW]
SLAP	^{2}H in water	$^{2}\delta = -428.0$ [vs SMOW]
PDB	^{18}O in CO_2	$^{18}\delta = 0$ [vs PDB]
$\cdots$	$\cdots$	$^{18}\delta = +30.6$ [vs SMOW]
PDB	^{13}C in CO_2	$^{13}\delta = 0$ [vs PDB]
NBS19	^{18}O in CO_2	$^{18}\delta = -2.20$ [vs PDB]
NBS19	^{13}C in CO_2	$^{13}\delta = +1.95$ [vs PDB]
N_2 in air	^{15}N	$^{15}\delta = 0$ [vs air Nitrogen]

Since δ is usually a small number, it is given in permil (‰), equivalent to a factor of 10^{-3}. Unless stated otherwise, the δ values of the water isotopes are given relative to the VSMOW standard.

During routine measurements, a reproducibility of $\sigma = \pm 0.1$‰ for the Oxygen-18 isotope and $\sigma = \pm 1$‰ in the case of Deuterium can be achieved.

A number of absolute abundance measurements of VSMOW were made, albeit at a precision which falls short of that of the differential measurements. Lately, with improvements in the absolute isotope ratio measurements by the Avogadro project, a measurement of the ^{18}O abundance at a reproducibility level of 10^{-5} is underway, based on O_2 as the measurement gas rather than the conventionally used CO_2, as described in Box 2.3.

Box 2.3. The absolute calibration of the delta scale.

In addition to the aesthetically unsatisfactory situation of having a standard (or standards) whose isotope abundance is inadequately known and which is inherently unstable in time (in the case of the water standard) or possibly not quite homogenous (in the case of a mineral standard), there are recurring practical problems that invite an absolute measurement of the isotope abundance. One of the problems relates to the wide range of isotopic values in the water cycle which requires the use of at least two reference standards to bracket the isotopic value of the measured samples, since a linear extrapolation of the mass-spectrometric measurements may not be appropriate and, actually differ from machine to machine. Indeed, two reference standards, namely VSMOW and SLAP which differ by about 55.5‰ in δ^{18}O, are provided by the IAEA. Unless a reliable calibration of both these reference standards is available, one is faced by two or more uncompatible δ–scales: one based on a linear interpolation between the standards and another based on an extrapolation of measurements from one standard assuming ideal linear MS performance. An even more serious problem arises when isotope ratios between different chemical species are to be compared. This applies, in particular, to the determination of isotopic fractionation between co-existing phases such as water and carbonates, silicates, phosphates, etc. For the differential ratio measurements it is then necessary to convert the oxygen in the different materials into a common gas to be measured, where every conversion procedure potentially introduces some isotope fractionation. Neither the equilibrium fractionation factors nor the fractionation introduced by incomplete conversions are adequately known. The common use of semi-empirical correction factors in these procedures introduces uncertainties that are one to two orders of magnitude larger than the reproducibility of the mass-spectrometric measurements. It is only an absolute measurement of the isotope ratio that would enable one to control the reliability of the conversion procedures, especially for the case when data from different laboratories and procedures are to be matched. The measurement of the absolute isotope ratio of ^{2}H/^{1}H in VSMOW was attempted by comparing these to a synthetic mixture

(Continued)

Box 2.3. (*Continued*)

of the isotopes, at a reproducibility of 2.10^4 (Hagemann *et al.*, 1970). For many other "reference materials" and other isotopic species there are no absolute isotope ratio determinations and their comparison to SMOW and other standards depends on estimates of the isotope fractionation between the different co-existing materials. However, recently, the isotope abundance measurements have been much improved and reproducibilities of 10^{-4} to 10^{-5} can be obtained for the measurement of absolute ratios of oxygen isotopes in O_2 gas. Since fluorination techniques can produce O_2 gas from water, carbonates, sulfates, silicates and phosphates at close to 100% yield, thus being free of fractionation effects, it has been suggested that atmospheric O_2 gas from areas of either sources or sinks of oxygen be used as a primary standard for the oxygen isotopes. Following the accurate calibration of such a standard by means of synthetic mixtures of the isotopes concerned, a comparison can then be made between CO_2 produced by conversion of the standard to CO_2 with the CO_2 produced in each case by the accepted analytical procedures, i.e. H_2O–CO_2 equilibration, acid decomposition of carbonates, etc (Gat and DeBievre, 2002).

2.2. The natural abundance of isotopes of oxygen and hydrogen

The two Boato diagrams (Figs. 2.1 and 2.2) show the range of the abundance of the heavy isotopes in natural materials (Boato, 1961). $\delta^{18}O$ values span a range of $\pm 50‰$ around SMOW, with most of this range pertaining to the waters of the hydrologic cycle. The range of δ^2H values is considerably larger. In the water cycle, the meteoric waters, i.e. the precipitation and water derived directly from precipitation are relatively depleted in the heavy isotopes, whereas surface water exposed to evaporation and water in geothermal systems fall on the enriched side of the scale.

The average of the present hydrosphere is estimated at $\delta(^{18}O) = -0.64‰$. The value of the primordial water, i.e. that water first condensed over the primitive earth and still possibly present deep down under the earth's surface, has been of great interest. The value suggested is $\delta^{18}O = +5‰$ and $\delta^2H = -55‰$ and is based on the isotopic composition of old rocks formed at high temperatures (so that no isotope fractionation can be expected) and on the value of water in magmatic exhalation. The evolution in the δ value of the hydrogen isotopes from the primordial to the present day hydrosphere is explained by the slow but preferential loss to space of ^1H atoms from the exosphere. In the case of the Oxygen-18, the loss of heavy oxygen can be accounted for by the deposition of carbonates

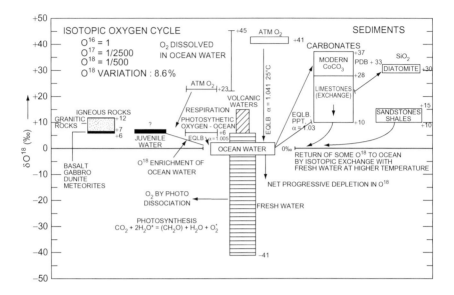

Fig. 2.1.

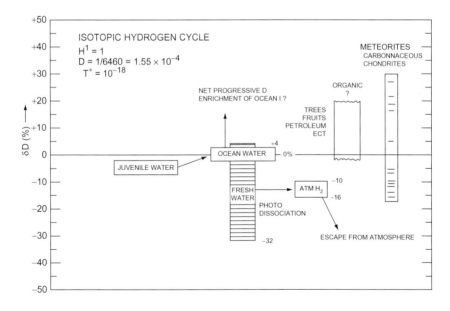

Fig. 2.2.

Box 2.4. Oxygen-17 Labelled Water.

Classical equilibrium isotope fractionation theory (Bigeleisen, 1962) predicts that the isotopic changes of the Oxygen-17 labelled molecules would be about one half that of the equivalent ones labelled by Oxygen-18. The prevalent measurement techniques for Oxygen-17, namely by NMR or Mass Spectrometry on CO_2 (where the molecule $^{13}C^{16}O_2$ of mass 45 masks the changes in abundance of the sought after molecule $^{12}C^{16}O^{17}O$), were not sensitive enough to detect any deviations from such a relationship within the hydrologic cycle. However, it was shown that extra-terrestrial abundances of ^{17}O in meteorites differed markedly from the expected ones (Clayton, 1993). Moreover, measurements of the isotopic composition of stratospheric oxygen and CO_2 (cf. Thiemans *et al.*, 1995) established an anomalously large enrichment of the ^{17}O-labelled molecules of oxygen and ozone; this was shown to be the result of a mass-independent fractionation by photochemical reactions in the upper atmosphere.

In order to exploit this isotopic tracer pulse for studies of the mixing and incorporation of the high altitude oxygen in the terrestrial geochemical cycles (Luz *et al.*, 1999), the Oxygen-17 analytic procedures had to be improved. This was achieved by using Oxygen gas for the measurement in the MS, which was produced from the water by electrolysis (Meijer and Li, 1998) or by a fluorination technique, with CoF_3 as the fluorination agent. Barkan and Luz (2005) perfected this procedure to obtain a precision of 0.01‰ to 0.03‰ for both $\delta^{17}O$ and $\delta^{18}O$. Using this method, they determined the fractionation factor $^{17}\alpha$ and $^{18}\alpha$ for the liquid/vapour equilibrium and found the ratio of $\ln^{17}\alpha/\ln^{18}\alpha$ to be constant with a value of 0.529 ± 0.001 over the temperature range of $11.4°$ to $41.5°C$. Moreover, a large suite of samples of meteoric waters followed a similar relationship. However, the relationship between the fractionation effects of the diffusion of the isotopic water molecules through air during the evaporation of water followed a smaller slope of 0.518, so that these two effects can be distinguished by measuring both the ^{17}O and ^{18}O isotopes. Examples of use of this methodology is found in Landais *et al.* (2008).

and other sedimentary minerals (which are relatively enriched in the heavy isotope) from the world ocean over the geologic time span.

The ^{17}O abundance is estimated at about 37.4×10^{-3}. By and large, the variability in ^{17}O mimics that of ^{18}O at about one half the amount of change, since most of the isotope fractionation occurs in a mass-dependent fashion. However, there is mounting evidence that some mass-independent fractionation takes place during photochemical interactions in the higher atmosphere. The case of the ^{17}O labeled waters is discussed in Box 2.4.

Chapter 3

Isotope Fractionation

Introduction

Due to the difference in the properties of the isotopic atoms and molecules, be it rates of motion, frequencies of inter-molecular vibrations, rotations or stability of chemical bonds, the relative abundance of the isotopes in the source material and the product of a dynamic process is different. Even in the so-called *equilibrium systems*, where the amount and concentration of the chemical compounds involved do not change, one encounters differences in the abundance of the isotopic species in the various components of such a system. The term *"Isotope Fractionation"* is used to denote any situation where changes in isotopic abundances result.

Based on the process which is responsible for the change, one distinguishes between *equilibrium* fractionation (also termed thermodynamic fractionation), and *kinetic* and *transport* fractionations. These will be discussed in the following sections. All of these are *mass-dependent* effects, in the sense that there is a quantitative difference in the rates or extent of the isotope fractionation which depends directly on the mass difference between the atoms or molecules involved. Where a number of isotopes of the same element are involved, such as the case of ^{18}O, ^{17}O, ^{16}O or ^{14}C, ^{13}C, ^{12}C, the mass dependency is reflected in the degree of enrichment or depletion of the isotope with intermediate mass, which is in between that of the heavier and lighter isotopes (Bigeleisen, 1962).

The mass dependency also explains the fact that some of the fractionation effects involving the hydrogen isotopes are larger by almost one order of magnitude when compared to those involving the oxygen or carbon isotopes because of the relatively large ratio in mass of 2:1 in the case of the hydrogen isotopes compared to 18:16 and 13:12, respectively, for the oxygen and carbon isotopes.

There are, however, two situations where *mass-independent* fractionation is encountered. The first situation is when a nuclear interaction occurs with one or the other of the isotopes being either the reactant or product. The other case is that of a photo-chemical reaction, whereby the bond involving just one of the isotopic species is preferentially excited by radiation of the appropriate wavelength.

Changes in the (averaged) isotope composition of different compartments or fluxes in the hydrological cycles can also be encountered without any fractionation when these compartments are not well mixed or equal in either time or space, and then partially utilized. Such a process, termed a *selection process* to distinguish it from a fractionating one, is one based for example on the seasonal changes in the isotope composition of atmospheric, surface or soil waters, as described in Chapter 8.

3.1. Isotope fractionation under equilibrium conditions (Thermodynamic fractionation processes)

Compounds A and B with a common element (with more than one isotopic species) which are in equilibrium with respect to the exchange of the isotopes between them, may have different isotopic compositions R(A) and R(B) resulting from differences in bond strength of the isotopic species. An isotopic fractionation factor is then defined as follows: $\alpha_{(A-B)} = R(A)/R(B)$. In analogy to the reversible chemical reactions at equilibrium (exchange reactions), one can define a thermodynamic constant of the isotopic exchange reaction:

$$AX^0 + BX^1 \leftrightarrow AX^1 + BX^0$$

where X^0 and X^1 stand for the two isotopes of the common element between the molecules concerned, as follows:

$$K_x = \frac{[AX^1] \cdot [BX^0]}{[AX^0] \cdot [BX^1]}. \tag{3.1}$$

The brackets [ ] signify the thermodynamic activity of the component concerned.

K_x equals $R(A)/R(B)$ to the extent that one can assume no isotope effect on the activity coefficients ratio for the two chemical species; this, however, is a first approximation at best.

The same principle can be applied also to the equilibrium between two phases of some material, such as water in its three phases: vapour, liquid and solid (ice). The special case of the relationship between the concentration and activity fractionation factors in the case of saline solutions was discussed in Box 2.2.

These equilibrium fractionation effects are dominated by the zero-point energies of the binding energies; being quantum mechanical effects they are appreciable at lower temperatures and practically disappear at higher temperatures, typically following a rule such as:

$$\log_e \alpha = C_1/T^2 + C_2/T + C_3 \tag{3.2}$$

where T is the temperature expressed as $^\circ$Kelvin.

For the case of the vapour to liquid equilibrium of water, the following relationships were determined by Majoube (1971):

$$\log_e {}^{18}\alpha_{L/V}: \quad C_1 = 1.137.10^3; \quad C_2 = -0.4156; \quad C_3 = -2.0667.10^{-3} \tag{3.3}$$

$$\log_e {}^{2}\alpha_{L/V}: \quad C_1 = 24.844.10^3; \quad C_2 = -76.248; \quad C_3 = +52.612.10^{-3}. \tag{3.4}$$

These data are shown in graphical form in Figure 3.1. The equilibrium isotope fractionation factors are often denoted as α^* or α^+, where by a convention introduced by Craig and Gordon (1965) one uses the notation of α^* when $\alpha^* < 1$ and α^+ for $\alpha^+ > 1$.

Table 3.1 summarizes some of the fractionation factors which operate during phase transitions of water. It is evident from the magnitude of the fractionation factors during the phase transitions of water, that the resultant changes in isotope composition of hydrogen and oxygen respectively, differ by almost a factor of 10. In order to portray these then in δ (‰)-space (i.e. the plot of ${}^2\delta$ versus ${}^{18}\delta$), one foreshortens the ${}^2\delta$ axis by this factor of 10 relative to the ${}^{18}\delta$ axis. The trend lines relating the isotopic composition of vapour to the liquid or solid at thermodynamic equilibrium are seen to be temperature-dependent, ranging in slope, S (where $S = \Delta^2\delta/\Delta^{18}\delta$), from S = 8.65 at 0°C to S = 7.75 at 20°C for the liquid/vapour, and S = 7.7 for the vapour/ice equilibrium at 0°C. Notably, the equivalent value for the solid/liquid transition is much less than these, being S = 5.86.

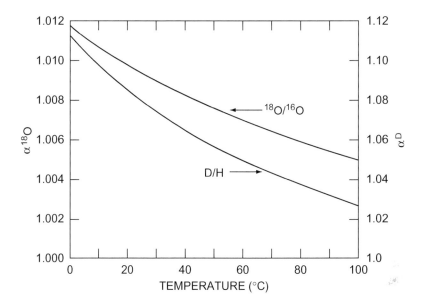

Fig. 3.1.   Isotopic Species of water.

Table 3.1.   Equilibrium isotope fractionation factors for phase transitions of water.

	$^{18}\alpha^{+}$	$^{2}\alpha^{+}$	Reference
Solid (ice)–vapour			
at $0°$C	1.0152	1.132	[1]
at $-10°$C	1.0166	1.151	[1]
Liquid–vapour			
at $+20°$C	1.0098	1.084	[2]
at $0°$C	1.0117	1.111	[2]
Solid (ice)–liquid			
at $0°$C	1.0035	1.0208	[1]/[3]
Hydration water–solution			
Gypsum ($CaSO_4 \cdot 2H_2O$)			
at $20°$C	1.004	0.985*	[4]/[5]

[1] Majoube, 1971a; [2] Majoube, 1971b; [3] Arnason, 1969; [4] Fontes and Gonfiantini, 1967; [5] Sofer, 1975.
*Value refers to $^{2}\alpha = $ R(water in gypsum)/R(mother solution).

3.2. Isotope fractionation accompanying transport processes

The effect of the mass difference of isotopic molecules is expressed most directly in their transport properties. In the hydrologic cycle, differences in the rate of diffusion of the water molecules through air is of major concern. Based on the gas-kinetic theory (Chapman and Couling, 1951), where the diffusivity of molecule A through gas B is expressed by

$$Đ_{A,B} = \frac{\{C \cdot (1/M_A + 1/M_B) \cdot T^{3/2}\}}{\{P \cdot \sigma^2 \cdot \Omega_{A,B}\}}$$

one would expect the following ratios of diffusivities through air for the deuterated and Oxygen-18 substituted water molecules $^1H^2HO$ and $H_2^{18}O$, respectively, relative to the common water molecule $^1H_2^{16}O$, (under neglect of the interactive term $\Omega_{A,B}$ and on the assumption of negligible change of the molecular volume with isotopic substitution):

$$Đ(^1H_2^{16}O)/Đ(^1H^2H^{16}O) = 1.015 \quad \text{and} \quad Đ(^1H_2^{16}O)/Đ(^1H_2^{18}O) = 1.030.$$

Contrary to this expectation, the experimentally measured values were found quite different (Merlivat, 1978) especially for the case of the deuterated molecules:

$$Đ(^1H_2^{16}O)/Đ(^1H^2H^{16}O) = 1.0250 \quad \text{and} \quad Đ(^1H_2^{16}O)/Đ(^1H_2^{18}O) = 1.0306.$$

This is attributed to interactions between water molecules in the gas phase. Such an effect is even more accentuated in the liquid phase. Indeed, Wang *et al.* (1953) measured appreciable differences between the diffusion coefficients of isotopic water molecules of equal mass in liquid water, for example, the molecules $^1H^3HO$ and $H_2^{18}O$. Other transport properties in the liquid, such as viscosity, are similarly affected by the intra-molecular forces and especially by hydrogen-bonding.

Lately, this apparent discrepancy in the Diffusion Constant of the deuterated water molecules in air has been questioned by Cappa *et al.* (2003), based on a re-evaluation of the isotope fractionation process during the evaporation of water, as will be discussed in more detail in Chapter 4.3. However, this subject is still not agreed upon and at this stage, the evidence seems to support the more classical approach based on Merlivat (1978).

Unlike the thermodynamic and kinetic isotope effects with their strong (negative) temperature dependence, the effect of temperature on the diffusion coefficients in air is rather weak, i.e. Đ is proportional to $\sqrt{T}$ and the ratio of diffusivities is practically invariant with respect to temperature.

In the early literature, the fractionation effects that result from differences in the movement of the isotopic molecules were termed *"kinetic isotope fractionation"* effects, a term which nowadays is reserved for the effects resulting from the breaking and making of chemical bonds in chemical and bio-chemical reactions. The presently discussed effects are termed *"transport isotope fractionation"*.

Chapter 4

Models of Isotopic Change
in the Water Cycle

4.1. Closed equilibrium system

In an assemblage of N water molecules consisting of ^iN and ^jN isotopologues where j refers to the more abundant isotope species, so that $R = {^iN}/{^jN}$ and $N = ({^iN} + {^jN})$, the distribution of the isotopologues between two coexisting phases of the water (for example between liquid and vapour or liquid water and ice) under equilibrium conditions is governed by the appropriate "Equilibrium Fractionation Factor". In the case that the two phases contain $N_{(1)}$ and $N_{(2)}$ molecules, respectively, then using a box-model formulation the distribution of the isotopes between the 2 phases, namely, the relation of $R_{(1)} = \{{^iN_{(1)}}/{^jN_{(1)}}\}$ to $R_{(2)} = \{{^iN_{(2)}}/{^jN_{(2)}}\}$, will be given at equilibrium by $\alpha = R_{(2)}/R_{(1)}$.

α is the relevant "Equilibrium Fractionation Factor" at the temperature concerned. For the case of the natural isotope abundances of either the Oxygen–18 or Deuterium, where $^jN \gg {^iN}$, one can write, to a good approximation, that $N = ({^iN} + {^jN}) \approx {^jN}$ and $d^jN \approx dN$.

Suppose that initially all the water is in one phase (N_0 and $^iN_0$ molecules, respectively) and that subsequently more and more of it changes its phase under equilibrium conditions. Such is the situation when fog develops as an air mass cools as well as when clouds form in an ascending air-mass. If $0 < f > 1$ is the fraction of water remaining in Box #1, i.e. $f = N_{f,(1)}/N$, then the evolution of the isotopic composition, described by the ratio R_f, can be described by the following equation provided the equilibrium situation is maintained between the two phases throughout; note that if the temperature changes throughout the process, the controlling fractionation factor changes accordingly.

$$R_f \cdot \alpha = \{{^iN_0} - {^iN_f}\}/\{N_0 - N_f\} = \{R_0 - f \cdot R_f\}/(1 - f) \qquad (4.1)$$

In δ (‰) nomenclature where $\varepsilon = (\alpha - 1)10^3$, the equivalent formulation is:

$$\delta_f \approx \delta_0 - \varepsilon \cdot (1 - f) \tag{4.2}$$

4.2. Open equilibrium systems — the Rayleigh Equation

When material is removed from a mixed assemblage of isotopic molecules in the form of another phase under momentary equilibrium conditions, such as when water droplets are removed from moist air or the evaporate flux from a liquid, then given the initial isotopic ratio in the mixed reservoir (box) of $R = {}^iN/{}^jN$, the isotopic ratio in the effluent flux, $R_{ef} = d^iN/d^jN$, is given by the relationship:

$$\alpha = R_{ef}/R = \{d^iN/d^jN\}/\{{}^iN/{}^jN\} \tag{4.3}$$

when α signifies the Equilibrium Fractionation Factor.

When more and more of the material is removed from the reservoir, the changing isotope composition is described by the "Rayleigh Equations." This formulation is based on one given by Lord Rayleigh (1902), who 100 years ago thus described the enrichment of alcohol by distillation from an alcohol/water mixture. Even though these equations were formulated to describe a momentary *equilibrium* situation, they can be applied also in the case that fractionation factors other than the "Equilibrium Fractionation Factor" operate.

One needs to distinguish between the following situations:

(1) Systems closed to inflow, open to outflow:

    A) where material is removed only according to Eq. 4.3
    B) where part of the outgoing flux occurs without any change in the isotope composition, e.g. in a bottom-leaking evaporation pond.

(2) Systems open to both inflow and outflow:

    C) balanced throughflow systems, i.e. influx equals efflux:

        C1) where material is removed only according to Eq. 4.3
        C2) where part of the outgoing flux occurs without isotope fractionation

    D) unbalanced systems

4.2.1. *Systems without inflow*

The changing isotope ratio as material is removed under equilibrium conditions, based on the relationship of Eq. 4.3 and the approximation discussed above, is then given by:

$$dR/dN = d(^iN/N)/dN = \{d^iN/dN - {}^iN/N\}/N = \{R/N\} \cdot (\alpha - 1) \quad (4.4)$$

from which it follows that:

$$d\log R/d\log N = (\alpha - 1) \tag{4.4a}$$

Defining $f = N_f/N_0$ as the fraction of material remaining in the system after part of it was removed (where N_0 is the number of molecules at the beginning of the process), then for a constant value of the fractionation factor, the evolution of the isotope ratio will be given by integration of Eq. 4.4 as follows:

$$R = R_0 \cdot f^{(\alpha-1)} \tag{4.5}$$

In $\delta(\text{‰})$ nomenclature, since $R/R_{standard} = \{1 + \delta/10^3\}$ we have

$$d\log R = d\log\{1 + \delta/10^3\} = \{d\delta/10^3\}/\{1 + \delta/10^3\}$$

so that

$$d(\delta/10^3)/d\log f = (\alpha - 1) \cdot \{1 + \delta/10^3\}$$

For small δ values this can be approximated by:

$$\delta \approx \delta_0 - (\alpha - 1) \cdot \log f \tag{4.6}$$

The change in isotope composition can thus be used as a measure of the degree of rainout or evaporation, as the case may be.

When only a part of the efflux occurs under fractionating conditions, the change of the isotope composition according to Eq. 4.6 will not be commensurate with the actual water balance. Comparison of the expected and actual isotopic change will reveal the partitioning of outflow by the fractionating and non-fractionating modes.

4.2.2. *Systems with both in- and outflow*

The mass balance of such a system with both in- and outflow is given by:

$$dN = dN_{in} - dN_{out}$$

When $dN_{in} = dN_{out}$ the system is at steady state relative to the amount of material. The isotope balance is then given by:

$$d(RN)_s = R_{in} \cdot dN_{in} - \alpha \cdot R_s \cdot dN_{out}$$

Such a system can reach an isotopic steady state when $R_{in} = \alpha \cdot R_s$.

Details of such systems will be further discussed in the context of the discussion on lakes.

Figure 4.1 shows the evolution of the isotopic composition for these different model systems. Evidently, the change in isotopic composition is quite different, even though all are governed by the same fractionation factor.

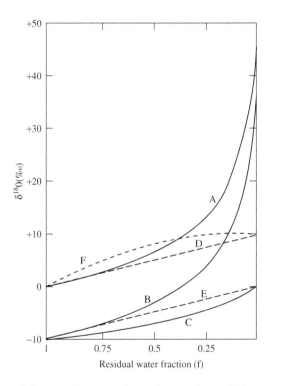

Fig. 4.1.   Change of the isotopic composition of water from which vapour is removed under conditions of equilibrium-fractionation for different scenarios: Curves A,B,C: under a Rayleigh regime, showing the residual water, the instantaneous vapor flux and the total removed vapour, respectively. Curves D,E: The liquid and vapour composition, respectively, for a closed system. Curve F: A liquid approaching steady state (horizontal axis in this case represents a time axis).

4.3. Evaporation under natural conditions

The early surveys of the isotope composition of terrestrial waters reported
an enrichment of the heavy isotopes in some lakes and evaporative basins.
This was first attributed to a *Rayleigh process*, based on the *Equilibrium
Isotope fractionation* between liquid water and its vapour. However, the
high enrichments in extreme evaporative systems which were thus expected
were not realized. This led to a series of experimental evaporation studies
in the field and laboratory, among others by Dansgaard (1954), Friedmann
et al. (1956), Craig Gordon & Horibe (1963), and Lloyd (1966). It was
thus realized that the isotopic buildup during the process of evaporation
into the atmosphere is not a simple one-way *"Rayleigh"* process, but is
limited by the back-diffusion of the atmospheric moisture. This was further
confirmed by observations on the Tritium balance of lakes (e.g. Oestlund
and Berry, 1970) where the vapour exchange between the surface waters
and the ambient air could be identified as an important factor.

More quantitative evaluations of the buildup of the heavy isotopes
(either of the oxygen or hydrogen isotopes) during evaporation further
suggested that the stable isotope fractionation in accompaniment of the
evaporation is larger than that to be expected based on the *"Equilibrium
Fractionation"* factor. The comparison of the buildup of the Deuterium with
that of the Oxygen-18, as shown by Craig (1961), also indicated that the
relative rate of change could not be explained by simply applying an *equilibrium fractionation* to the evaporation process. To account for this discrepancy, a number of possible *"transport"* effects were considered, namely:
the diffusion of water molecules through the air column, self-diffusion in
the liquid or the rates of detachment or attachment of water molecules
at the liquid surface. The fact that the diffusion of the water molecules
through the air column is an important factor in the isotope fractionation
that accompanies the evaporation process was established by comparative
laboratory experiments in a nitrogen and helium atmosphere, respectively,
as reported by Gat and Craig (1965).

With these findings in mind, and based on the realisation that the rate
of evaporation of water into un-saturated air above it is limited by the
transport of vapour from the layer immediately above the surface into the
ambient atmosphere (Brutsaert, 1965), Craig and Gordon in 1965 suggested
the model schematically shown in Fig. 4.2 to describe the isotope fractionation during the evaporation process; it follows the formulation of the
Langmuir linear-resistance model (Sverdrup, 1951).

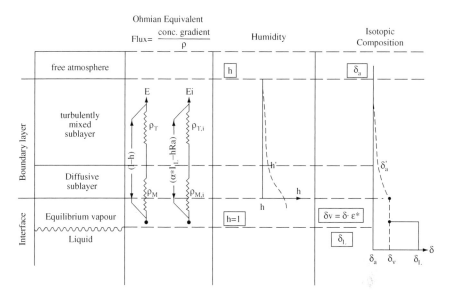

Fig. 4.2. The Craig–Gordon model of isotope fractionation during the evaporation from a free water surface.

The following assumptions underlie this model:

- The establishment of liquid-vapour equilibrium at the water-air interface is rapid compared to the vapour transport from the interface zone to the free ambient atmosphere, so that isotopic equilibrium between the surface water and the saturated vapour above it can be assumed. The (saturation) vapour content and the appropriate isotope fractionation factor thus depends on the temperature and salinity of the water surface (*cf.* Chapter 2 and Box 2.2).
- The vertical flux is the balance between the upward evaporation flux and the back-diffusion of the vapour from the ambient atmosphere.
- No divergence or convergence in the vertical air column.
- No isotope fractionation in accompaniment of fully turbulent transport.

In the model, the vapour flux is described in terms equivalent to *Ohm's Law* as the quotient of the concentration gradient (expressed in terms of the humidity gradient, normalized to a value of $h = 1$ in the saturated surface layer) to the transport resistance, ρ. In the case of a stagnant air layer above the liquid surface, as pertains to the evaporation through an unsaturated soil layer as well as the water loss from plants through the stomata (to be described below), the vapour transport is by molecular diffusion and a fixed linear concentration profile is established. The water and isotope fluxes through the air column are then determined by their respective molecular diffusion coefficients. On the other hand, for an open interface under strong wind conditions, most of the transport is by turbulent diffusion, and molecular diffusion through a non-steady variable air layer is effective only close to the surface (Brutsaert, 1965).

The appropriate flux equation for the water substance is then as follows:

$$E = (1 - h)/\rho \qquad (4.7)$$

where ρ is the sum of the resistance to diffusive flow in the diffusive sub-layer, ρ_M, and in the turbulently mixed sub-layer, ρ_T, namely:

$$\rho = \rho_M + \rho_T.$$

The corresponding flux equations for the water isotopologues, either of $^1H^2HO$ or $H_2^{18}O$, are:

$$E_i = (\alpha \cdot R_L - h \cdot R_A)/\rho_i \qquad (4.8)$$

where $\rho_i = \rho_{i,M} + \rho_{i,T}$.

The isotope composition of the evaporation flux, R_E, is then the quotient of Eqs. 4.8 and 4.7:

$$R_E = E_i/E = [\alpha \cdot R_L - h \cdot R_A]/[(1 - h) \cdot (\rho_i/\rho)] \qquad (4.9)$$

This can be expressed in the more conventional δ-units as follows:

$$\delta_E = [\alpha \cdot \delta_L - h \cdot \delta_A - \varepsilon^* - \Delta\varepsilon]/[(1 - h) + \Delta\varepsilon/10^3]$$
$$\approx [\delta_L - h \cdot \delta_A - \varepsilon^* - \Delta\varepsilon]/(1 - h)$$

In this expression, $\varepsilon^* \equiv (1 - \alpha) \cdot 10^3$ and $\Delta\varepsilon \equiv (1 - h) \cdot (\rho_i/\rho - 1) \cdot 10^3$. In the linear resistance model, $\rho_i/\rho = (\rho_{i,M} + \rho_{i,T})/(\rho_M + \rho_T)$ so that the term $(\rho_i/\rho - 1)$ can be written as follows:

$$(\rho_i/\rho - 1) = (\rho_M/\rho) \cdot (\rho_{i,M}/\rho_M - 1) + (\rho_T/\rho) \cdot (\rho_{i,T}/\rho_T - 1).$$

The term $(\rho_{i,T}/\rho_T - 1)$ can be eliminated, based on the model assumption that there is no isotope fractionation during turbulent transport so that $\rho_{i,T} = \rho_T$. Defining $\theta = (\rho_M/\rho)$, the expression for $\Delta\varepsilon$ is then:

$$\Delta\varepsilon = (1 - h) \cdot (\rho_M/\rho) \cdot (\rho_{i,M}/\rho_M - 1) \cdot 10^3$$
$$= (1 - h) \cdot \theta \cdot (\rho_{i,M}/\rho_M - 1) \cdot 10^3 \qquad (4.10)$$

In the case of a stagnant diffusion layer, ρ_M is proportional to $Đ^{-1}$. Under strong turbulent wind conditions and a rough water surface, the transient-eddy model of Brutsaert (1975) can be applied where ρ_M is proportional to $Đ^{-1/2}$. For more moderate interface conditions, a transition from the proportionality of $Đ^{-1/2}$ to $Đ^{-1}$ occurs (Merlivat and Coantiac, 1975). As described in Chapter 3.2, the ratio of the molecular diffusivities through air of the water isotopologues differ appreciably from those expected from the gas-kinetic theory, based on the molecular weights only. Defining the ratio of the molecular diffusivities of the heavy and light water molecules, respectively, as C_K then, the expression for $\Delta\varepsilon$ becomes:

$$\Delta\varepsilon = (1 - h) \cdot \theta \cdot n \cdot C_K. \tag{4.10a}$$

The value of n ranges between the value of $n = 1$ for stagnant, and $n = 1/2$ for fully turbulent wind conditions, respectively.

The formulations just presented apply to a well-mixed and homogenous water body where at all times there is no difference of isotope composition between the surface and the bulk of the liquid. This is not necessarily true in the case of open water under wind-still conditions or when evaporation takes place from within the soil. Under such circumstances, a concentration gradient can be established in the liquid boundary layer which slows down the isotopic change that accompanies the evaporation, as elaborated in Box 4.1. The evaporation from within the soil column will be discussed in Chapter 8.

Due to the additional transport-fractionation imposed on the evaporation process under natural conditions, the ratio of change of the Deuterium to Oxygen-18 labeled waters during such a process differs from that of the liquid-vapour transition under equilibrium conditions. On a plot of $^2\delta$ versus $^{18}\delta$ (δ-*plot* for short), such evaporated waters are then located on "Evaporation Lines", whose slopes are reduced relative to the Equilibrium Lines. These relationships will be further elaborated in the context of the discussion on surface waters in Section 9.4 and Fig. 9.8.

The historical evolution of the concepts leading to the Craig-Gordon Evaporation Model has recently been reviewed by Gat (2008) and the model was re-evaluated by Horita, Rozanski and Cohen (2008).

Box 4.1 The case of incomplete mixing in the liquid surface layer.

The formulations of the Craig–Gordon model shown were based on the assumption that the evaporating water body was well mixed, so that at all times, $\delta_L(\text{surface}) = \delta_L(\text{bulk})$. This is not necessarily so for an open water body under no-wind conditions or for evaporation from within a porous medium, e.g. soils. Incomplete mixing can then result in the enrichment of the surface water in the heavy isotopes relative to the bulk of the liquid and, hence, the establishment of a concentration gradient in the liquid boundary layer. In the Craig–Gordon model, one then has to introduce an additional resistance, ρ_L.

On the assumption of a surface boundary layer of constant depth, the following formulation represents the isotope flux through the boundary layer:

$$E_i = E \cdot R_L - (R_{\text{surface}} - R_L)/\rho_L \qquad (1)$$

On adding this expression to the flux equation in the atmospheric boundary layer, where the term $R_{surface}$ substitutes for R_L and disregarding at this stage any change in the values of α^* and ε^* due to a temperature or salinity change in the surface layer, then the increase of the isotope composition resulting from the evaporation is given by Eq. 2:

$$d\ln R_L/d\ln N = [h \cdot (R_L - R_a)/R_L - \varepsilon^* - \Delta\varepsilon]/[(1-h) + \Delta\varepsilon + \alpha^* \cdot E \cdot \rho_L] \quad (2)$$

to be compared to Eq. 3 for the mixed surface waters:

$$d\ln R_L/d\ln N = [h \cdot (R_L - R_a)/R_L - \varepsilon^* - \Delta\varepsilon]/[(1-h) + \Delta\varepsilon]. \qquad (3)$$

Obviously then, when $\rho_L \neq 0$, the isotopic change is slowed down.

4.3.1. *Isotope composition of the evaporation flux* — δ_E

The isotopic composition of the evaporation flux is usually the least known parameter in the balance equation of the water substance, as it is not directly amenable to measurement. Based on the Craig–Gordon model of evaporation from an open-water body, it is expressed by Eqs. 4.9a and 4.10. For operational use, this can be written in the following form:

$$\delta_E \approx [\delta_L - h \cdot \delta_A - \varepsilon^* - (1-h) \cdot \theta \cdot n \cdot C_k]/[1-h] \qquad (4.11)$$

where $C_k = (\rho_{i,M}/\rho_M - 1) \cdot 10^3$. Based on the measurements of Merlivat (1978) this has the value of $C_k = 28.5(‰)$ in the case of the oxygen-18 labelled water molecule and $C_k = 25.1(‰)$ for the deuterated water molecule diffusing through air; n is the exponent in the relationship between ρ_M and D, which, as discussed above, is usually assumed to be $n = 1/2$ for evaporation from the ocean. $\theta = (\rho_M/\rho)$; this can be assumed equal to 1 for a small water body whose evaporation flux does not perturb the ambient moisture significantly. However, it was shown to be $\theta = 0.88$ in the case of the North American Great Lakes (Gat *et al.*, 1994) and that a value of $\theta = 0.5$ applies in the eastern Mediterranean Sea (Gat *et al.*, 1996). The latter value appears to be the limiting factor encountered also over large open oceanic areas.

4.3.2. *Postscript*

As mentioned in Section 3.2, Cappa *et al.* (2003) from recent measurements on the isotope composition of evaporating waters concluded that the main reason for the apparent discrepancy between the expected and measured fractionation due to diffusion of the water vapour from the surface was not due to the Diffusion Constant (Đ), as given by Merlivat (1978), but due to a lower temperature at the water surface resulting from incomplete mixing of the water in the liquid boundary. This issue is not yet settled as of now, and an authorative re-evaluation of the isotope fractionation in accompaniment of evaporation is called for.

4.4. The "Isotope Transfer Function" (ITF)

The set of rules that describe the change in isotope composition resulting from the passage of the water substance from input to output of a system is termed the *Isotope Transfer Function* (Gat, 1997). Two effects have to be taken into account namely, on the one hand, those actuated by fractionation between the isotopes in accompaniment of phase transitions or kinetic and transport processes and, on the other hand, when there is a selection of part of the input waters during the transition to the output, provided these selected parcels are characterised by a different isotopic composition. An apparent further change can occur simply by mixing of water parcels of different composition. An example is the change of isotope composition that accompanies the transition of precipitation input through the land/biosphere/atmosphere interface to groundwater recharge flux or surface runoff, as discussed in more detail in Chapters 8 and 9.

Figure 4.3 summarizes the occurrence of these different processes throughout the water cycle.

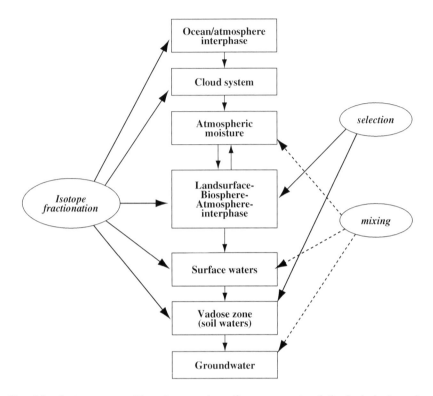

Fig. 4.3.   Isotope composition changes along the components of the hydrologic cycle.
The diagram indicates the sites where isotope fractionation, selection or mixing processes
operate.

Chapter 5

The Ocean System and the Marine Atmosphere

5.1. Isotope composition of the ocean waters

Compared to the rest of the hydrologic cycle, the isotope composition of the ocean waters in the large oceanic basins is fairly homogenous, with values ranging just about $\pm 0.5\permil$ around SMOW in the case of $\delta(^{18}O)$ and a factor of 10 larger in the case of $\delta(^2H)$. This was established in early surveys by Redfield and Friedman (1965), Epstein and Mayeda (1953), Dansgaard (1960) and Craig and Gordon (1965) and led to the suggestion of establishing the *Mean Ocean Water*'s isotope composition as the reference for the δ-scale of isotopic composition in the water cycle (Craig, 1961). Somewhat larger variations are on record along the continental margins of the ocean, especially in bays or semi-enclosed marine basins (as reviewed by Anati and Gat, 1989) as well as at the ice-covered margins of the ocean in the Arctic and Antarctic area (reviewed by Ferronsky and Brezgunov, 1989).

The variations in the isotopic composition are generally correlated with the salinity variations in the oceans, even though somewhat different correlations apply in different areas of the ocean. Surface waters in the trade-wind regions are somewhat enriched in the heavy isotopic species along with a salinity increase, reflecting the negative water balance between evaporation and precipitation, $(E/P) > 1$. In contrast, in regions affected by the freezing and melting of seawater, the salinity change is accompanied by only a minor isotopic change, due to the relatively small isotopic fractionation for the liquid/ice phase transition (*cf.* Table 3.1). However, the melting of the continental ice sheet, where the ice is the result of accumulation of isotopically very depleted precipitation, imparts a marked change on the isotope composition in parallel with the reduction of salinity. This phenomenon of the melting of the continental ice shield is indeed believed to be at the core of the change in the ocean water's isotope composition between glacial and interglacial periods, (termed the *glacial increment*) estimated to be of the order of $-0.5\%_0$ and $+1.0\%_0$ in $\delta(^{18}O)$ around the present value of SMOW for the full interglacial and glacial-maximum periods, respectively. The estimate of the corresponding change of the hydrogen isotope composition is based on the average composition of the icecap, which is estimated to be situated on the *Global Meteoric Water Line* with a value of close to $\delta(^{18}O) = -30\%_0$ and $\delta(^2H) = -230\%_0$.

The relationships discussed above are shown schematically in Fig. 5.1. The distinctive isotopic compositions of the surface waters in the different geographic settings thus provide a tracer for the formation and mixing characteristics of the deep waters in the various marine basins, as discussed in detail by Craig and Gordon (1965).

In enclosed and marginal sea basins, where the hydrological balance is dictated as much by the interaction with the continental environment as by the water exchange with the open oceanic source, the isotopic composition strikes a balance between that of the ocean water, evaporation and the dilution by freshwater influx of precipitation, river and groundwater discharge and of melt-waters. One distinguishes between predominantly

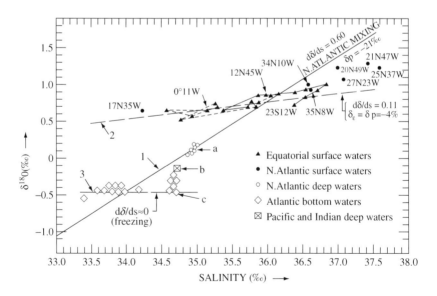

Fig. 5.1.   The isotope/salinity relationship in representative marine waters. (Based on data reported by Craig and Gordon).

evaporative basins such as the Red Sea (Craig, 1966) and the Persian Gulf (Lloyd, 1966; McKenzie *et al.*, 1980) in which the waters are enriched in the heavy isotopic species alongside a salinity increase and systems such as Baltic (Ehhalt, 1969; Froehlich *et al.*, 1988) and the Black Sea (Swart, 1991) as well as numerous estuarine bays where the freshwater inputs with their depleted isotope values result in isotopic compositions which are a mixture of the fresh and marine waters; since the isotope composition of the fresh influx waters is very varied and depends on their origin and hydrological history with a strong seasonality, each such system portrays a different set of isotope-salinity relationships (Friedman *et al.*, 1964), as exemplified in Fig. 5.2.

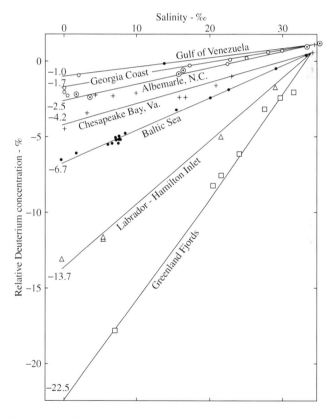

Fig. 5.2. The relationship between the depletion in Deuterium (measured as δ^2H(%) relative to SMOW) and the dilution of the marine salinity in estuaries and marginal inland seas. (note that δ^2H is given in percent, rather than in permil units, as is the accepted mode today).

The over-riding control on the degree of deviation of the isotopic composition of these marine systems from that of the adjoining oceanic water mass is, however, exerted by the coupling and mixing modes with the ocean; these range from stratified countercurrent fluxes over restricting straits such as in the strait of Gibraltar connecting the Mediterranean Sea to the Atlantic Ocean, the Skagerrak which links the Baltic Sea to the North Sea or the straits at Bab-el-Mandeb linking the Red Sea and the Indian Ocean, to open estuarine bays where mixing is primarily controlled by the ocean currents and the tidal regime.

The largest land-enclosed marine basin is the Mediterranean Sea, which extends over a number of climate regions and is influenced both by evaporative enrichment (the Mediterranean is an important source of moisture for adjacent land regions, primarily during the winter months) but also, on the other hand, is the recipient of drainage from large continental areas, most prominent among them the Danube and Volga draining through the Black Sea, the Rhone River in the west and the Nile River draining into the south-eastern part. As a result, recorded isotope values range from $\delta(^{18}O) = -2.5‰$ at the Black Sea (Swart, 1991), to values as high as $\delta(^{18}O) = +2.5‰$ in the eastern, more arid parts of the Mediterranean Sea. Moreover, the relative variation of the Oxygen-18 and Deuterium values, which in other marine systems are well correlated (as shown in Fig. 5.2), are more complex in the Mediterranean where actually the increase in $\delta(^{18}O)$ is not accompanied by a commensurate change of $\delta(^2H)$. This was explained by Gat *et al.* (1996), as shown in Fig. 5.3, by the interplay of evaporative enrichment and freshwater dilution.

The noticeable variations of the isotope composition and of salinity throughout the marine domain obviously provide a useful tracer to oceanic mixing processes. A recent compilation of published data on the isotopic composition of sea waters by Schmidt *et al.* (1999) ranges from $\delta(^{18}O) = -6‰$ to $\delta(^{18}O) = +3‰$. The whole set can be downloaded from URL://www.giss.nasa.gov/data/o18data/. The more depleted values are found near the poles where the input of meltwater into the ocean accounts for relatively large local deviations.

5.2. Water in the marine atmosphere

5.2.1. *Isotopic composition of marine vapour*

Measurements of the isotopic composition of water vapour above the ocean, performed at deck or mast height during cruises in the Atlantic, Pacific and

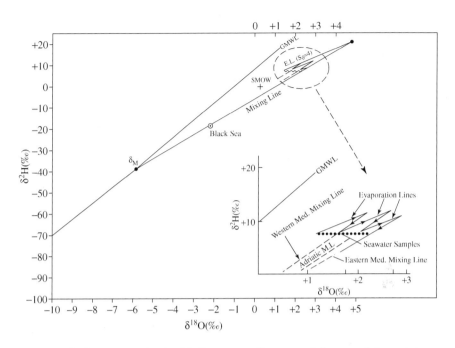

Fig. 5.3. Surface waters in the Mediterranean Sea on a δ-diagram. Interpreted as a balance between the evaporative enrichment of the surface waters (along "Evaporation Lines") and their dilution by freshwaters (along "Mixing Lines"). δ_M signifies the assumed meteoric water input (situated on the GMWL) which determines the mixing line for the eastern part of the sea.

Indian oceans during the 1960s, showed values of $\delta(^{18}O)$ between $-11‰$ and $-14‰$, with corresponding deuterium values that fit rather closely to the GMWL relationship of $\delta(^2H) = 8 \cdot \delta(^{18}O) + 10‰$. Since at the prevailing temperatures, the expected equilibrium vapour composition would range from about $\delta(^{18}O) = -9‰$ to $-8‰$ for the range of the surface water isotope compositions of $\delta(^{18}O) = 0‰$ to $+1‰$, it is evident that the ambient vapour above the ocean is not the *equilibrium vapour*. Neither does it match the isotope composition of the evaporation flux, δ_E, which was estimated to be on the average $\delta(^{18}O) \approx -4‰$ based on material balance consideration between the evaporation and the precipitation in the marine domain, on the assumption of a closed hydrological cycle over the ocean. In order to account for this discrepancy, Craig and Gordon (1965) proposed a model of the evaporation, mixing and precipitation over the sea, shown in Fig. 5.4, which explains in general terms how the evaporation flux is mixed in the atmospheric boundary layer with the residual vapour from

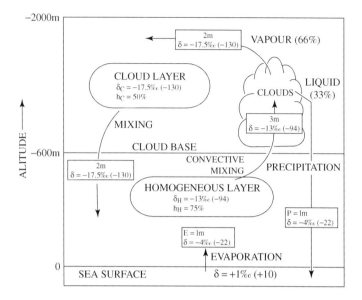

Fig. 5.4.   The marine-atmosphere isotope model of Craig and Gordon, assuming a steady-state closed system. The first δ-value signifies the δ^{18}O value and the δ^2H value is given in parentheses.

aloft, following condensation and rainout of an amount of water (1 m of liquid water depth on a global average), which is the equivalent of the 1 m of water evaporated from the oceans, *i.e.* disregarding the 8% of the marine moisture which is advected onto the continents (*cf.* Fig. 1.1). A companion model that also includes the terrestrial water cycle (under a steady state assumption) and is thus closer to verisimilitude, is shown in Fig. 5.5. A more detailed discussion of the estimate of the evaporation flux's isotope composition, based on the Craig-Gordon evaporation model, was presented in Chapter 4.

The assumption of evaporation taking place into an atmosphere of 75% humidity (*cf.* Fig. 5.5) accounts for the *d-excess* value of 10‰ in the atmospheric moisture, that conforms to the worldwide average of isotopic labeling of meteoric waters (Craig, 1961). This is illustrated on a δ-scale diagram in Fig. 5.6. Based on a similar "world-wide-averaged marine model" applied to the conditions of the glacial epoch, as proposed by Merlivat and Jouzel (1979), the meteoric waters were characterized by a lower *d-excess* value of around $d = 6$‰. This corresponds to a lower humidity deficit over the oceans at that time, namely h $\approx$ 80%.

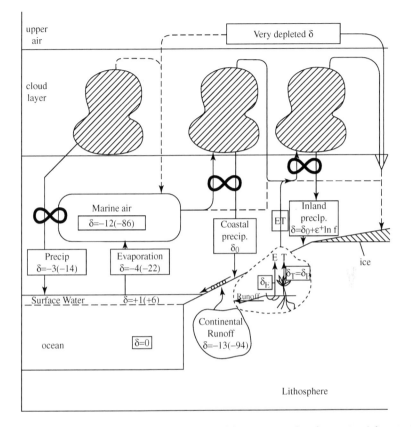

Fig. 5.5. Adaptation of the Craig-Gordon model to account for the terrestrial part of the water cycle. The double loop signifies that the precipitation and vapour fluxes are in isotopic equilibrium.

In more detail, the isotopic composition of marine vapour and the resulting marine precipitation is much more variable than would be expected based on these simplified models. This is the result, foremost, due to the different humidity and temperature over parts of the ocean surface. Whereas in the major oceanic source region for moisture, the atmospheric waters are indeed characterized by the *d-excess* value of close to 10‰, values closer to d = 13‰ represent the moisture over the southern ocean (giving rise to the precipitation characterized by such values over parts of the Australian continent and in South America). In the colder parts of the North Atlantic and Arctic oceans values of d < 10‰ apply, whereas high values (of up to d = 35‰) occur in sea areas on the lee-side of continents

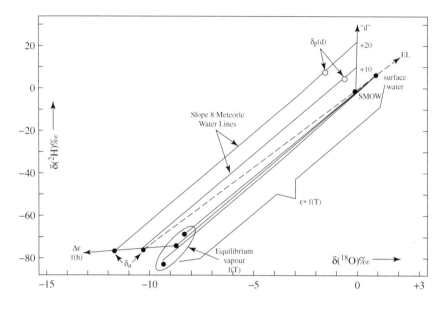

Fig. 5.6.   The marine evaporation process in terms of the δ-scale.

such as the Mediterranean (Gat *et al.*, 2003), the China Sea (Liu, 1984) or
the Atlantic sea coast of North America.

Because of the insufficient data base, a detailed representation of the
isotopic composition of the marine moisture in time and space is still very
sketchy. It is obvious, however, that an interplay of a number of factors is
involved, among them:

• The surface water's isotope composition;
• The humidity over the ocean;
• the physical structure of the interface and, in particular, the presence of
  foam (white-cap conditions) and the possibility of spray formation;
• the presence of either convective or subsiding air motions;
• the hydro-meteorological history of the air masses aloft.

Relating to the third of these processes, the contribution of complete
or partial evaporation of spray droplets to the evaporation flux would
introduce the liquid without or with little isotope fractionation into the
atmosphere, thus constituting a bypass to the process described by the
Craig-Gordon model, resulting in lower *d-excess* values than expected
according to the model. Since a large portion of the evaporation over

the ocean occurs under strong wind conditions, this mechanism cannot be ignored. From the comparison of the measured isotope composition of vapour collected over the Mediterranean Sea in winter with the composition expected for the admixture of evaporated flux, Gat *et al.* (2003) estimated that up to one half of the added moisture was at times contributed by the spray-droplet pathway. Similar conclusions were drawn in the case of moisture contributing to tropical hurricanes by Lawrence *et al.* (1998).

The fourth process enumerated above has a profound effect on the humidity gradient over the ocean and thus on the isotopic signature of the atmospheric waters. Under a regime of subsidence, the vertical mixing is minimal, so that the air close to the seawater/air interface can almost reach saturation conditions, characterized by a low *d-excess* value. Such a situation typically occurs in the Mediterranean Sea during summer; it is usually not conducive to the occurrence of precipitation with the exception of *orographic* rains formed when these marine air masses are lifted during the passage over coastal or island mountain ranges. In contrast, under convergent conditions, fresh and relatively dry air masses are continuously introduced close to the sea surface, resulting in enhanced evaporation under conditions of a humidity deficit, thus resulting in a larger *d-excess*.

The air at the upper limit of the atmospheric boundary layer is fashioned by the hydro-meteorological history of the air masses at this altitude, that can be determined by a *Lagrangian* back-tracking procedure (Wernli and Davies, 1997). Its isotope composition and water content then determine the moisture and isotope gradient in the boundary layer.

Three basic models have been applied to describe the evolution of the marine atmosphere, as an initially "dry" air moves over the ocean, picking up more and more moisture (Fig. 5.7); the vertical stability of the air layers at the site of evaporation appears to be a dominant factor, interacting with the large scale moisture transport at higher altitudes. Further afield, the local evaporation-precipitation balance as described by the large scale marine models becomes decisive.

5.2.2. *Precipitation over the oceans*

One would expect the isotopic composition of marine precipitation to be close to that of a first condensate of the marine vapour, in isotopic equilibrium with it at the appropriate temperature, with values ranging typically from $\delta(^{18}O) = -1.5\%$ to $\delta(^{18}O) = -5\%$. This is indeed the case, generally, but unlike the precipitation over continental areas where the

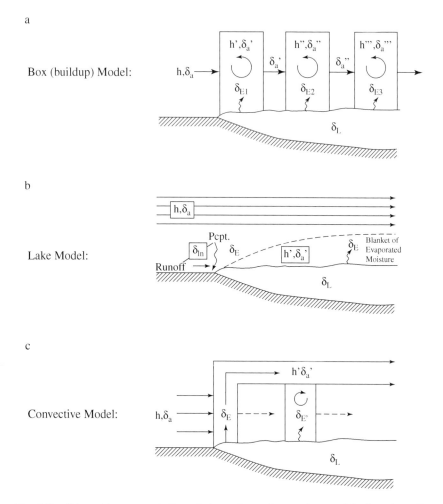

Fig. 5.7. Schematic representation of three air-sea interaction models (based on Gat *et al.*, 2003).

changes in the isotopic composition of both the hydrogen and oxygen are well constrained along the relevant MWL, the marine precipitation, samples of which are shown in Fig. 5.8 based on the GNIP data, show a relatively wide scatter of the *d-excess* parameter, apparently representing the locally different evaporation conditions over the ocean. These variations are then averaged and smoothed out when the marine air is disengaged from its vapour source as it moves onto the continents.

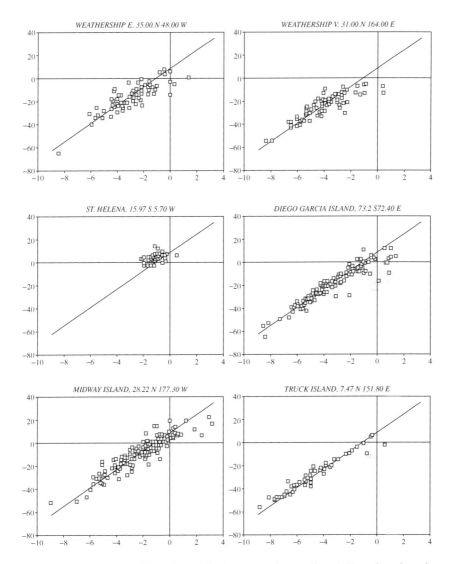

Fig. 5.8. Isotope composition of precipitation at marine weather stations, based on the GNIP data set.

There are some exceptions to the rules discussed where one observes rain in the marine domain which is more depleted in the heavy isotopes. Most notable are the rains in the tropical regions that are associated with the towering clouds of the Inter Tropical Convergence Zone (ITCZ). As an example, near the Brazilian coast as well as in the Pacific Ocean region, rains are encountered with isotopic values as low as $\delta(^{18}O) = -10\%_0$ (Matsui *et al.*, 1983), values which are more typical of mid-continental locations.

The general pattern has been aptly summarized as follows (*vid.* Section 4.3 in Gat, Mook and Meijer, 2001):

"In precipitation over the ocean, collected at island stations or weather-ships, the range of $\delta(^{18}O)$ values is relatively small, with but little seasonal changes in many cases and a lack of clear correlation with temperature (Rozanski *et al.*, 1993). There is a relatively large variability in the value of the *d-excess* in the case of oceanic precipitation, especially close to the major source regions of the atmospheric moisture. On plotting the isotopic data of both oxygen and hydrogen (the δ-plot), the scatter in the *d-excess* predominates, whereas the isotope data from inland stations even though they show a much larger range of $\delta(^{18}O)$ values, are well-aligned along meteoric water lines and the scatter of the *d-excess* is actually reduced. This characteristic can be quantified by recording the ratio of the spread of the *d-excess*, $\sigma(d)$, to the range of the $\delta(^{18}O)$ values, i.e. $\sigma(d)/\langle\delta(^{18}O)\rangle$. As an illustration, this value changes from 2.0, 2.3, 1.8 and 2.8 at the coastal and island stations of the European continent at Valentia (Ireland), Reykjavik, Faro (Portugal) and Weathership E (north Atlantic) to values of 0.4, 0.35 and 0.7 at continental stations such as Berlin, Krakow and Vienna, respectively."

Chapter 6

Clouds and Precipitation

Due to the decreasing temperature with altitude, any air mass containing water vapour that rises under adiabatic conditions will reach an altitude where the air is saturated relative to its moisture content (the dew-point) and condensation to droplets can commence. As the ascent continues, more and more of the moisture will condense, resulting in growth of the droplets and their fall to the ground as rain. Under colder conditions and at higher altitudes in the atmosphere the freezing point is reached, so that solid precipitation elements (either snow, graupel or hail) are formed.

Precipitation is generated when warm and humid air masses

- slide up onto colder air-masses resulting in *frontal* rainfall
- updraft vertically, cooling adiabatically and condensing rapidly, forming rain or hail (*convective* rainfall)
- are forced to ascend along higher altitude landscape, thus cooling and forming the so-called orographic rain.

During the fall of the raindrops from the cloud base to the ground, the precipitation elements may undergo some evaporation, which under conditions of low air humidity can result in the total loss of the precipitation.

Frontal rainfall has the property of widespread occurrence; *orographic* rainfall is often local. *Convective* rainfall tends to be heavy and of short duration; it occurs frequently in tropical zones, less frequently so in other climates. During many rain events, the intensities often change from low ones at the beginning of the shower to high ones in the middle, petering out to low intensities towards the end of the rain event.

Rain intensities, frequencies and spatial distributions are rather variable in arid and semi-arid regions and to some extent in the tropics. They are less variable under polar, cold and temperate climates. Thus it is more difficult to quantify the hydrological inputs in the arid and semi-arid areas than in other climate zones, and the errors in the determination of groundwater recharge increases. One finds that in temperate and polar regions, good averaging of inputs would require observations over a period of 10 years, whereas as much as 20 years may be necessary in the arid, semi-arid or tropical regions. Due to the changing climate, no meaningful averaging is then possible.

Precipitation is measured at a given point by rain samplers and in integrated form over larger areas by means of calibrated weather radars, and more recently, by satellite imagery. The precision of such radar measurements for rain or snow, respectively, are 15% and 40%. In arid, semi-arid and tropical climate as well as in areas characterized by high relief, one needs more measurement points in order to evaluate the temporal and areal variability of the precipitation.

6.1. The isotopic composition of the precipitation

Under an idealised situation, when condensation takes place under equilibrium conditions without reaching super-saturation and when rain falls as soon as the droplets are formed, then the isotopic composition of the evolving rain shower would follow an ideal *Rayleigh Law* for an "Open Equilibrium System" (*v.* Chapter 4) corresponding to the appropriate temperature. On the δ-diagram, this follows rather closely along a *Meteoric Water Line* of slope: $\Delta\delta(^2\mathrm{H})/\Delta\delta(^{18}\mathrm{O}) \approx 8$ with slight variations depending on the temperature. In reality, however, a large proportion of the liquid remains suspended as floating droplets in the cloud, rapidly attaining isotopic equilibrium with the vapour phase. In this case, the isotopic composition of the evolving rain will be governed by the rules of the closed-equilibrium system (*v.* Chapter 4.1) becoming depleted or enriched in the heavy isotopes as the liquid water content in the clouds increases or dissipates, as shown in Fig. 6.1. Being a process also dominated by the equilibrium fractionation factors, the *d-excess* value, however, remains essentially invariant.

Whereas liquid droplets undergo molecular exchange with the ambient moisture (Box 6.1), this is not the case with solid precipitation elements. These preserve the isotopic composition acquired at the point of their formation, often under conditions very different from those at the cloud base or the ground, to be described in Section 6.1.1.

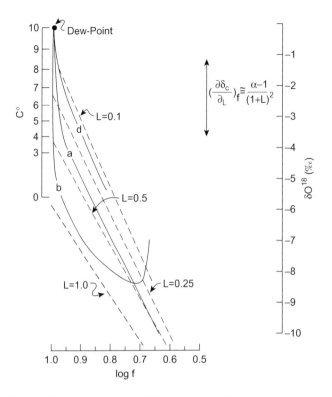

Fig. 6.1. Change of the isotopic composition of precipitation as a result of rainout for an air mass with a dew point of 10°C (based on Gat and Dansgaard, 1972). f-the remaining fraction of the atmospheric moisture. The solid lines give the δ values in precipitation for the classical Rayleigh process. The lines marked L refer to precipitation according to the 2-phase model (L being the liquid water content in the air mass). The dashed line shows the composition of the liquid phase in a closed system without rainout; (in this case f measures the fraction remaining in the vapour phase). The thick curved line (b) represents the evolution of the isotopic composition of precipitation for the passage over a mountain chain, accompanied first by buildup and then dissipation of the liquid water content in the clouds.

Box 6.1. The interaction between rain droplets and the ambient moisture.

The exchange of water molecules between a liquid drop and the ambient vapour results in the establishment of isotopic equilibrium in those cases where the air is saturated with respect to the liquid at the prevailing temperature. Evaporation or condensation of water occurs when the air is under — or oversaturated, respectively, with respect to saturated vapour pressure. This is accompanied by the isotope fractionation characteristic of these

(Continued)

Box 6.1. (*Continued*)

processes. Once equilibrium has been established, a dynamic exchange of water molecules continues without however, leading to a visible change.

The kinetics of the exchange process involves three steps: a very rapid process at the very surface of the liquid controlled by gas kinetics and much slower processes of mixing into the air and bulk of the liquid, which are controlled by the convective currents and turbulence in both these media. The situation of a falling drop through the air, accompanied by currents engendered by this, enhances the efficiency of both of the latter processes.

The overall kinetics of the exchange process, like all exchange processes, follows first-order kinetics and can thus be characterized by a half-life or a relaxation time, i.e. the time to achieve $1/e$ of the final equilibrium state. For the case of falling droplets, the size of the drop is a determining parameter, as it affects both the speed of fall and the size of the water reservoir. For raindrops falling at terminal free-fall velocities (at $10°C$), the following data, given by Bolin (1959) and Friedman *et al.* (1962), can serve as guidelines:

Drop radius (cm)	Relaxation distance (m)	Relaxation time (sec)
0.01	5.1	7.1
0.05	370	92.5
0.10	1600	246

A "*box model*" representation, while possibly adequate for describing the situation in a fog, is not suitable for the complex situation in most cloud systems. The closest to such a situation is encountered in rain produced over a warm front when the precipitating warm air mass gradually climbs upwards over the near-surface colder air cushion, as described by Bleeker *et al.* (1966), and also in *orographic* rains when the near-surface air layers ascend an elevated terrain. In convective clouds accompanied by strong vertical motions, a vertical gradient of isotope depletion is established in the ascending air mass in accompaniment of the decreasing temperature, and as more and more of the vapour is converted to the liquid (or solid) phases. At the top of large convective clouds, extremely negative δ-values have been measured; values as low as $\delta(^2H) = -600‰$ are reported in the upper troposphere (Rozanski, 2005, based on Taylor, 1972 and Ehhalt, 1971).

The isotopic composition of the precipitation has been found, on the whole, to be close to isotopic equilibrium with the ambient air at ground-level and strongly correlated with the temperature there (Yurtsever, 1975). The correlation is further improved when the cloud-base temperature is substituted for the ground temperature (Rindsberger and Magaritz, 1983). How this is compatible with the very depleted isotopic values encountered at the site of formation of the precipitation elements aloft, is explained by the molecular exchange between the falling rain drops and the ambient air (vid. Box 6.1). However, the details of the process depend on the aerodynamic structure within and around the cloud and in particular, the relative motion of the falling air droplets and the surrounding air. Three typical cases can be considered, namely a convective cloud wherein the rain falls counter-currently to the ascending air just like in a vertical "distillation column"; a leaning cumulus nimbus cloud under strong horizontal wind conditions, and the situation encountered in huge tropical clouds under the influence of the ITCZ as well as in tropical hurricanes where the updraft on the margins of the system is compensated by downdraft in its center, with the result that the rain drops fall within descending air in contact with vapour whose isotope composition is also very depleted in the heavy isotopic species. For example, at Belem at the mouth of the Amazon River, Matsui et al. (1983) encountered isotopic values as low as $\delta(^{18}O) = -10\%$ in rain at ground-level during such conditions, whereas otherwise rain in that locality is close to equilibrium with the marine-air moisture, as expected.

As mentioned above, the exchange process is not effective as long as the precipitation elements are solid and it begins to be actuated only below the freezing level. Another process which affects the isotopic composition of the falling raindrops is the partial evaporation of the drops, which takes place beneath the cloud base where the air is unsaturated with respect to its moisture content, resulting in enrichment of the heavy isotopes in the residual droplet. Since this enrichment of the heavy isotopes during evaporation occurs along *Evaporation Lines* with slopes less than 8 in δ-space, these evaporated waters are characterized by *d-excess* values that are lower than that of the original rain droplets, whereas the evaporated moisture shows the opposite effect. The evolving rain which develops when this near-surface air is then incorporated into the cloud is described in Box 6.2. Similar relationships, notably the increase of the *d-excess* in the atmospheric waters, result when the evaporation flux from surface water is incorporated into a precipitating system, as first shown over the African continent by Dansgaard (1964).

Box 6.2. Modification of the atmospheric moisture's isotopic composition by incorporation of evaporated moisture from rain droplets during their fall beneath the cloud base.

The schematic illustration, adopted from Gat (2000), portrays the processes occurring during a rain-shower beneath the cloud base as the falling droplets are partially evaporated and the evaporated moisture then mixes with the ambient air and is swept up into the cloud as the rainout continues. Assuming an initial isotopic composition of $\delta_{a,0}$ of the water vapour, this yields rain of composition $\delta_{p,0}$ at the cloud base. Evaporation results in enrichment of the heavy isotopic species in the remnant drop along *Evaporation Lines*. Since the slope of these lines in δ-space is usually less than that of the Meteoric Water Lines, the *d-excess value* of the resultant rain at ground level, whose isotopic composition is δ_1, is less than that of the pristine moisture whereas the recycled evaporated moisture's *d-excess* value is larger. When the evaporate then mixes into the ambient air to produce the "second generation" air moisture, etc., their isotopic composition ($\delta_{a,1}$, $\delta_{a,2}$) as well as that of the resulting precipitation ($\delta_{p,1}$) is characterized by a higher *d-excess* value than that of the first rain.

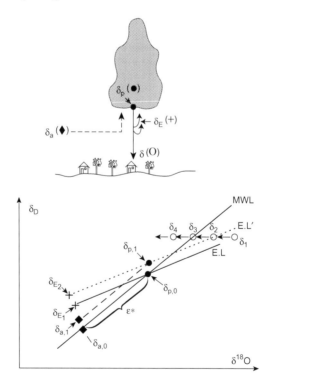

Fig. A. Modification of the atmospheric moisture's isotopic composition by incorporation of evaporated moisture from rain droplets during their fall beneath the cloud base.

It should be noted, however, that the *d-excess* value is not exactly conserved at most temperatures between liquid and vapour even under equilibrium conditions, because of the temperature dependence of the isotope fractionation factors of the hydrogen and oxygen isotopes. Especially at temperatures close to the freezing point, where the ratio of enrichments $\varepsilon(^2\text{H})/\varepsilon(^{18}\text{O})$ is as high as 9.5, the location relative to the MWL differs somewhat for the liquid and vapour samples even under equilibrium conditions.

6.1.1. *Snow, hail and other solid phase precipitation*

Based on the equilibrium isotope fractionation factors between water vapour and ice (*vid.* Table 3.1), one could expect solid precipitation elements to be more enriched in the heavy isotopes than liquid formed under the same conditions and with a somewhat lower *d-excess* value. Unlike liquid cloud droplets that change their composition by continuous exchange with the ambient moisture (and thus reach equilibrium even when initially this is not the case), the isotopic composition of the solid phases is frozen in. Indeed, detailed measurements of the isotope composition in hail particles, performed by Macklin *et al.* (1970) and by Jouzel *et al.* (1975), show the changing isotope composition as the growing hailstone ascends and falls within the cloud, which is preserved within the structure of the hail.

As discussed by Jouzel and Merlivat (1984), among others, it is observed that in many cases, the isotope composition of solid precipitation elements is characterized by a higher *d-excess* value than liquid precipitation. This was explained to be the result of the higher ratio of the ice/vapour fractionation factor for Deuterium compared to Oxygen-18, respectively, when a Rayleigh Model is applied to the removal of the precipitation from the cloud. However, comparison of measured to calculated values based on such a model necessitated the introduction of a correction factor which was attributed by Jouzel and Merlivat to an additional process, namely a *transport fractionation* during the diffusion of super-saturated water vapour onto the solid surface. In other cases where the solid is the result of the freezing of up-drafting liquid droplets, as is typical for the so-called *graupel*, the isotopic composition corresponds to that of the liquid in the lower part of the cloud. Probably the closest that one approaches a true equilibrium situation is in the case of *rimed pellets* and *hoarfrost*. Rather surprisingly, in spite of the varied isotopic signatures of the different pathways of formation of solid precipitation elements, the basic altitude, inland and temperature effects on the isotopic composition of precipitation are on the whole also observed in the cold regions.

6.1.2. *Precipitation over the continents*

As the marine air-masses invade the continents, labeled by their respective *d-excess* values, they are disconnected from their vapour source so that any precipitation reduces the moisture content and changes the isotopic composition of the remaining atmospheric moisture, unless compensated by evapo-transpiration from the land surface (to be discussed in more detail in Chapter 8). On the whole, the change in the isotope composition as more and more of the advected moisture is precipitated can be described by a Rayleigh relationship. Dansgaard (1964) in the first analysis of the global network of isotopes in precipitation (GNIP) suggested four "effects" which operate, namely the latitude effect, the inland (distance from the coast) effect, the altitude effect, and the amount effect. All except the last are strongly correlated with the decreasing temperature and have a marked seasonal component.

As can be seen from the world-wide compilation of the GNIP data (Fig. 6.2), the *latitude effect* is about $\Delta\delta(^{18}O) = -0.6‰$ per degree of latitude for continental stations in Europe and north America, and up to $\Delta\delta(^{18}O) = -2‰$ in the colder Antarctic continent. The *inland effect* varies considerably from area to area and from one season to another, even over a low-relief profile. During passage over Europe, for example, from the Irish coast to the Ural mountains, an average depletion of $\Delta\delta(^{18}O) = -7‰$ is observed; however, the effect in summer is only about one fourth of that in winter. This is attributed in addition to the difference in the temperature regime on the re-evaporation of much of the summer precipitation (Eichler, 1964). An extreme case of an absence of the inland effect over thousands of kilometers, in spite of strong rainfalls *en-route*, was reported over the Amazon basin by Salati *et al.* (1979). This is also attributed to the return flux of the moisture by plant transpiration, which essentially is a non-fractionation process (*vid.* Chapter 8) and thus invalidates the effect of the rainout; however, some of the return flux apparently also occurs by evaporation from open waters with the result that there is an increase of the *d-excess* in the atmospheric moisture. Such an increase is indeed noted in the comparison of data from the coastal and inland sites in the Amazon basin.

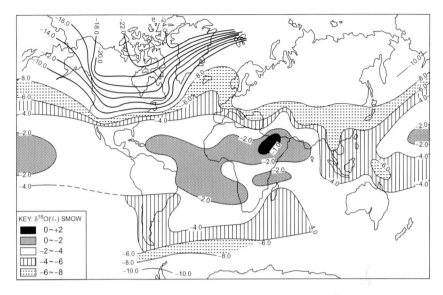

Fig. 6.2. Worldwide distribution of the annual mean values of $\delta^{18}O$ in precipitation, based on the data of the IAEA-GNIP network.

As a rule, the isotopic composition of precipitation changes with the altitude of the terrain, being more and more depleted in the heavy isotopes ^{18}O and ^{2}H at higher elevations. This has enabled one of the most useful applications in *Isotope Hydrology*, namely the identification of the elevation at which groundwater recharge takes place. The depletion of the heavy oxygen isotopes with elevation ranges between 0.1–0.5‰ per 100 meters and primarily results from the cooling of the air masses as they ascend a mountain, accompanied by the rainout of the excess moisture. A simple Rayleigh formulation indeed predicts altitude effects of this magnitude: for example, assuming an initial condition of marine air with a humidity of 80% at 25°C and a lapse rate of 0.53°C/100m, the effect is calculated to be about 0.25‰ per 100 meters. However other factors which change the isotope composition besides the basic Rayleigh effect need to be considered. One is the evaporative enrichment of isotopes in raindrops during the fall beneath the cloud base, as described in Box 6.1, which is larger at low altitudes where the cloud base is typically high above ground level. This so-called *pseudo-altitude effect* (Moser and Stichler, 1971) is observed clearly in inter-mountain valleys and on the leeside of a mountain range. The evaporative enrichment of the heavy isotopic species, unlike the primary Rayleigh

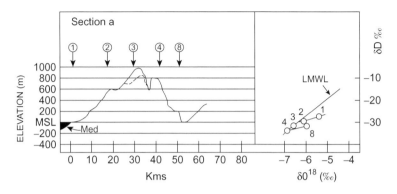

Fig. 6.3. The evolution of the isotopic composition of precipitation during the passage from the Mediterranean Sea over the Gallilean Mountains to the Rift Valley. Adapted from Gat and Dansgaard, 1972.

rainout effect, also results in a decrease of the *d-excess* parameter and thus marks these situations clearly. This effect is illustrated for the case of a traverse across the Gallilean mountain range and in the distribution of the *d-excess* parameter throughout the area, which closely follows the topography (Fig. 6.3 from Gat and Dansgaard, 1972). The most elusive factor is when different air masses with different source characteristics affect the precipitation at the base and crest of a mountain. A prominent case is that of the western slopes of the Andes in South America: precipitation near the crest results predominantly from air from the Atlantic with a long continental trajectory whereas air from the Pacific Ocean, with predominantly oceanic attributes, affects the precipitation in the lower elevations. Under such conditions, one can encounter anomalously large altitude effects.

All in all, there is clear dependence of the mean isotopic composition on temperature, even though this relationship differs in different geographic areas. Gourcy *et al.* (2005) give values of $\Delta\delta(^{18}O) = 0.82\%_0$ per °C for cold locations with mean annual temperatures below -2°C and $\Delta\delta(^{18}O) = 0.53\%_0$ per °C at locations within the temperature range of 0°C to 20°C. At tropical stations, the temperature dependence is much less defined. This pattern evidently reflects the fact that both the latitude and altitude effects are the result of the temperature decrease actuating more condensation of vapour. The much larger scatter in this dependence at many stations when individual rain events are considered is due to the important control on the isotopic signal which is also exerted by the origin and trajectory of the relevant air-masses, especially so in areas where the vapour source for precipitation is very diverse; for example, in the American mid-West where air from the Pacific, the Canadian shield and the Gulf of Mexico produce part of the annual precipitation, or the southern European region which is under the influence of air masses of either Atlantic or Mediterranean origin.

Usually, the isotope composition undergoes a yearly cycle (Fig. 6.4 gives some typical examples) with the largest amplitudes in mid-continental and high latitude locations. Not only are the δ-values involved but also, though to a lesser extent, the value of the *d-excess*. As shown by Rozanski *et al.* (1993) for the European stations, the d value in winter is somewhat higher

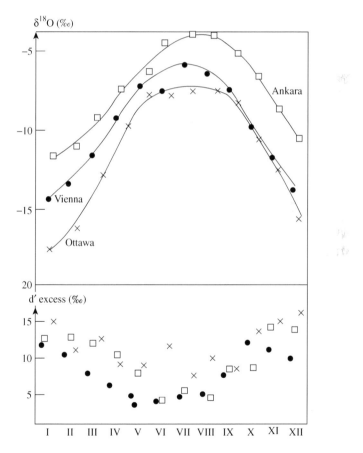

Fig. 6.4. Seasonal cycle of the isotope composition of the monthly-averaged precipitation at three representative continental GNIP stations.

than during summer. This seasonal variation has a number of contributions, and only to a small degree, that of the temperature effect. It may partially be the result of evaporation from the falling droplets in summer (resulting in lowering of the *d-excess* values) and the Jouzel-Merlivat effect during snow formation in winter, which imposes the opposite sense on this parameter. Moreover, a different source region for the precipitation during winter and summer could also be involved; a prominent example being the far-Eastern countries of Taiwan or Japan, as described by Liu (1984).

Rather surprisingly, there is a large variability in isotope composition on both an intra- and inter-rainspell scale (Pionke and DeWalle, 1992; Rindsberger *et al.*, 1990; Gat *et al.*, 2001). The variability between different rain events is caused predominantly by differences in the origin and rainout history of the air parcels concerned (for example *vid.* Legui *et al.*, 1983). The changes of the isotope composition during a rain event are a combination of the effect of the rainout history of the air mass and the amount effect. Often, the pattern of the evolution in time is that of a V or W shape, with the most depleted values corresponding to the core of the storm, as exemplified in Fig. 6.5. One notes that the range of δ-values over the duration of a single storm can be as large as that of the seasonal cycle.

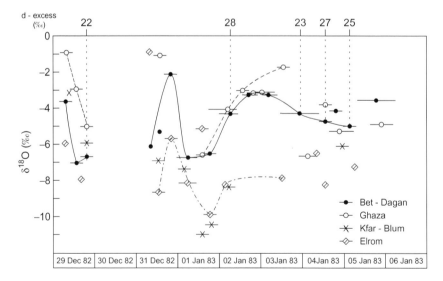

Fig. 6.5.   Intra-storm variations of the δ^{18}O value of precipitation during rain-spells at various stations in Israel and neighbouring locations.

In the vicinity of some oceanic areas where the presence of fogs is prevalent, fog drip and deposition is a noticeable component of the precipitation input. This is then usually characterised by an isotopic composition commensurate with the marine moisture and thus distinguishable from the precipitation, as it is relatively enriched in the heavy isotopes. Examples from the western coast of America (Arevena *et al.*, 1989; Ingraham and Matthews, 1995) from Pacific Islands (Scholl *et al.*, 2002; Scholl *et al.*, 2007) and from Africa (Ingraham and Matthews, 1988) have been described.

6.1.3. *The Local Meteoric Water Lines (LMWL)*

As we learned before, the isotope composition of precipitation at any specific location undergoes variations on an event-based, seasonal and inter-annual scale. These variations are the result of a number of factors, foremost the origin and rainout history of the precipitating air-mass, the temperature and, in particular, whether the precipitation consists of rain or snow, the rain intensity and cloud structure as well as the degree of evaporation during the descent of the rain droplets to the ground beneath the cloud base. Whereas on the whole, the precipitation δ-values are aligned along the Meteoric Water Line, there are deviations from strict adherence to this, so that the best-linear fit of the precipitation data in δ-space [the Local Meteoric Water Line- LMWL: $\delta(^2H) = a \cdot \delta(^{18}O) + b$] will show values of "a" which differ from the classical value of 8, in most cases a < 8.

Some examples of such LMWLs, constructed on the basis of monthly averaged values of the GNIP data set, are shown in Figs. 6.6(a–c). In Fig. 6.6a, for the data of Flagstaff, Arizona, the controlling reason for the relatively low slope is the partial evaporation of the falling droplets in this relatively dry climate, so that the enriched δ-values appear with a lower *d-excess* value. In contrast, in colder regions, as exemplified by the data from Ottawa, Canada (Fig. 6.6b), the winter snow samples with very depleted δ-values show higher *d-excess* values due to the Jouzel-Merlivat effect during snow formation. In stations where there are major changes in the origin of precipitation during the seasonal cycle, as in Tokyo, Japan (Fig. 6.6c), the LMWL can also deviate considerably from the GMWL.

The LMWL serves a useful purpose in defining the isotopic composition of the input into hydrological systems that are fed by meteoric water. In order to assess then, the possible changes introduced by additional processes during the passage through the land/atmosphere interface, the *d-excess*

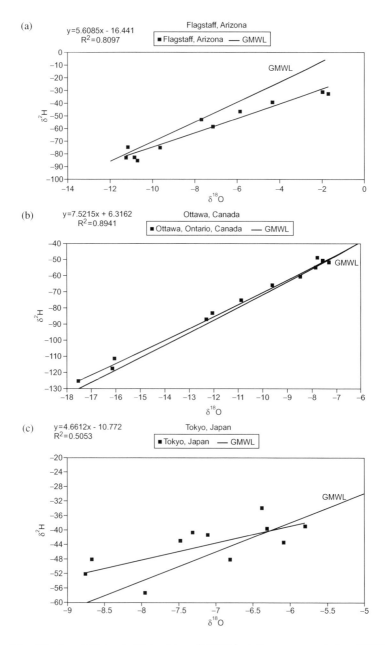

Fig. 6.6. The Local Meteoric Water Lines (LMWL) at three representative GNIP stations: A) Flagstaff, Arizona; a semi-arid climate location; B) Ottawa, Ontario; a cold-region station with snowfall during winter months; C) Tokyo, Japan.

value could be a misleading parameter. Rather the relative deviation of the isotopic composition of the waters concerned from the LMWL is a tell-tale parameter. This was defined as the "Line Conditioned Excess" parameter, LCE (Landwehr and Coplen, 2006) as follows: LCE $= \delta(^2H) - a \cdot \delta(^{18}O) - b$.

6.2. The isotope composition of the atmospheric moisture

Compared to the data set of the isotopic composition of precipitation, such as from the GNIP program, there is only very sketchy information on the isotopic composition of atmospheric moisture. The main reason is the difficulty in sampling the moisture without isotope fractionation, which requires rather elaborate techniques of freezing out water from large volumes of air under very controlled conditions. Such techniques are not routinely activated. It is only in recent years that the direct measurement of the isotopic composition of water vapour by laser spectroscopy is coming into vogue, reviewed recently by Kerstel and Meijer (2005), so far at lower sensitivity than the classical MS measurements. Direct measurements in the upper troposphere from the NASA Aura spacecraft, reported by Worden *et al.* (2007), enabled a more quantitative approach to the tropical water cycle. Further widespread use of this tool is expected to increase the utilization of stable isotopes as an auxiliary tool in the assessment of the atmospheric water balance.

By default, the isotope composition of the vapour is often estimated by proxy, assuming isotopic equilibrium with the local precipitation. While this may be acceptable during rainy periods in a continental setting, as shown by Craig and Horibe (1967) and Jacob and Sonntag (1991), it is not the case in the vicinity of an evaporative source such as a lake (e.g. Fontes and Gonfiantini, 1967) or in coastal areas, and obviously not during prolonged dry periods. It should be noticed that when applicable, the equilibrium value must be calculated for the appropriate cloud-base temperature, and that under this condition, the isotopic composition of vapour and liquid is not necessarily related exactly by a MWL with a slope of 8, except when the temperature is about 22°C. Under most circumstances, the data of the vapour samples is situated slightly above the MWL, as defined by the precipitation, i.e. characterised by a higher *d-excess* value, a fact often erroneously interpreted as due to the admixture of evaporated (fractionated) moisture into the atmospheric water pool, originating either from the evaporating falling droplets below the cloud base or from evaporating waters at the surface. In colder regions, the sublimation of falling snow near the surface has been proposed as an alternative mechanism, considering that snow is often characterised by a higher value of the *d-excess* (*vid.* Section 6.1.1). It must be realized, however, that the precipitation of snow with a high *d-excess* value necessarily leaves the atmospheric waters with a deficiency of this parameter.

As the value of the atmospheric moisture's isotope composition is required for the proper assessment of the water balance of evaporating systems, a number of initiatives for obtaining a more comprehensive data set are in progress, among them the MIBA network (Moisture Isotopes in the Biosphere and Atmosphere) of the IAEA. The widespread use of the laser technology is expected to add to this endeavour.

Chapter 7

Snow and Snowmelt Processes

7.1. Solid precipitation elements

As briefly outlined in Section 6.1.1, the isotopic composition of solid precipitation elements varies widely, depending on the mode and location of formation. Unlike the liquid precipitation whose isotopic composition is adjusted to equilibrium with the ambient moisture by continuous molecular exchange as the raindrops fall to the ground, in the case of solid precipitation the isotopic composition as formed initially is essentially frozen in. This was shown in some pilot studies on hailstones by Bailey *et al.* (1969), Macklin *et al.* (1970) and Jouzel *et al.* (1975 and 1985) where by differential analysis of successive layers in the hailstones, the upward and downward path of the growing hailstone within the cloud could be traced by their isotopic signature.

Among the different forms of solid precipitation elements, one distinguishes between snow flakes, graupel, hailstones and rimed snow. These differ in the location and mode of formation within the cloud system and therefore often are characterized by different isotopic signatures. As described in numerous publications (e.g. Sugimoto *et al* 1988, 1989) *graupel* is made up essentially by the freezing of super-cooled updrafting water droplets and subsequent further ice deposition in the cold cloud layers; their isotope composition is then based mainly on the regular liquid/vapour equilibrium fractionation close to the freezing point and thus located on the extension of the relevant Meteoric water Line for the appropriate vapour source. Snow or *rimed* particles which are formed in the cold part of the cloud would be expected , on the other hand, to follow a line based on the solid/vapour equilibrium value, which portrays a line of higher slope on the δ-plot. In practice in many cases the snow is characterized by isotopic

values with a much higher than the expected *d-excess* value based on the
equilibrium fractionation factors. This was explained by Jouzel and Merlivat (1984) by an isotopic transport fractionation effect at snow formation
as a result of the fact that vapour deposition occurs under supersaturated
conditions.

Due to the situation described above, the so-called *Dansgaard effects*
(*vid.* Section 6.1.2) are less pronounced under snowfall conditions, even
though the pattern of dependence on temperature and continentality
can still be recognized (for example over north America as reported by
Friedmann *et al.* 1964).

7.2. Snow accumulation and changes in the snowpack[*]

As a rule, the falling snow accumulates on the top of the foregoing precipitation so that a more or less continuous depth record of the sequential precipitation events is established in the snowpack; most notably the seasonal
cycle can be well recognised. A major disturbing factor is caused by snow-
drift from adjacent areas, especially from higher altitudes where the snow
may be more depleted in the heavy isotopes than the local precipitation.

Following the deposition, one distinguishes between a dry-snow, a percolating and a soaked-snowpack situation. In the first case, the temperature
remains so low throughout, so that no melting occurs even in summer.
However, some redistribution and homogeneization of the isotopic profile
takes place through vapour transport, complementary to the roundening
of the ice pellets at the expense of rimed snowflakes. When some melting
occurs on the surface and this meltwater then percolates into deeper layers
where it refreezes, this further modifies and mixes the depth profile in the
accumulated snow. Depending on the details of the temperature regime,
this can result in a variety of effects. When some evaporation of the surface-
melted water occurs, then the evaporative signature (i.e. enriched isotopic
and decreased d-excess values) is disseminated throughout the snowpack.
To the extent that the partial melting and refreezing occurs under conditions of solid to liquid isotopic equilibrium, then the percolation flux is initially depleted in the heavy isotopes (leaving an enriched surface ice layer),
thus carrying the depleted waters downward into the snowpack. However,
when the top layer is completely melted (thereby adopting the isotope

---

[*]Compare review of Arnason (1981).

composition of the snow cover), then partial refreezing of the percolation flux will result in an ice layer at depth which is enriched in the heavy isotopes relative to the initial composition. Evidently, the detailed evolution is determined by the thermal regime imposed on the system and in particular, the sequences of the melting and freezing processes.

7.3. The snowmelt process

Snowmelt is a major component of the runoff flux in colder and high-altitude regions, and many attempts have been made to use its isotope signature as a tool for quantifying its contribution to the runoff in a hydrograph. However, due to the rather complex redistribution processes during maturation of the snowpack, the snowmelt waters are not characterised simply by the isotope composition of the snowfall. Different isotope signatures are imposed depending on whether the meltdown occurs on the margins of the snowfield or on its top (in which case it often records the depleted isotope signature of the snow), or when the bottom is mobilized by heating from the rock layer up. As more and more of the snowpack then melts in spring, the runoff's isotope composition becomes increasingly enriched in the heavy isotopes.

An interesting situation, reported by Sugimoto *et al.* (2003), develops in the permafrost regions; there the runoff during the snowmelt period consists primarily of the melted snow accumulation, whereas when the soil is not frozen, the runoff consists of a mixture of summer and snow water, which are mixed in the upper soil layer.

Chapter 8

The Land-Biosphere-Atmosphere Interface

The incoming precipitation is partitioned at, or near, the land surface into fluxes of surface runoff, infiltration into the ground and eventual groundwater recharge and a back-flux into the atmosphere by means of either direct evaporation from surface-exposed waters or by transpiration of the plant cover. This partitioning is governed, primarily, by the structure and nature of the surface, its morphological and ecological makeup and the land use pattern and, on the other hand, by the climate and, in particular, the precipitation regime, especially when it involves frozen or snow-covered regions. The antecedent moisture content at the surface is another important factor to take into account.

As precipitation falls to the ground on an initially dry surface, the incoming rain begins to fill the surface reservoirs. One can view the surface system as a structure of successive interception reservoirs such as the canopy, the ground-surface storage depressions, the topsoil etc., as was shown schematically in Fig. 1.4. Each reservoir has a threshold capacity so that drainage to the other reservoirs in-line commences only when this is satisfied. Eventually the reservoirs are incapable of accepting additional inputs and further rainfall will result in flow towards a surface drainage system. The accommodation of continuing rain then depends on the rate of providing space for additional water input, either through infiltration and percolation into deeper soil layers, or interflow. It has to be realised that the hydraulic properties of some of the reservoirs, e.g. the topsoil, are time dependent and change with the moisture history of the system.

During the interval between rain events, the reservoirs lose water by direct evaporation or continuing drainage and by means of water uptake by the roots and transpiration; the latter can affect deeper lying soil layers which are not otherwise prone to evaporative water loss. When rain starts anew, the partially emptied reservoirs are replenished, reactivating the processes of drainage, infiltration and runoff. The new incoming water then interacts with the antecedent water residues in a variety of ways, ranging from complete mixing (the bucket model, Manabe, 1969) to a piston-flow displacement (Gvirtzman and Magaritz, 1990) that applies primarily to the flow in soil capillaries. One needs to consider further the possible heterogeneity of the surface and soil layer, with "by-pass" fluxes occurring along fissures and selective conduits, characterized by smaller threshold and holdup volumes than the *"matrix flowpaths."*

While the rainfall continues, the processes are essentially controlled by its amount and intensity relative to the holding capacity of the surface reservoirs and the rate of infiltration. The surface system's response time is commensurate with the precipitation event, except in the case of snowfall and frost conditions when long delays are imposed until melting sets in. As rain ceases, evapo-transpiration is the dominant water balance process, possibly interrupted by dew deposition or fog-drip (Ninari and Berliner, 2002). Depending on the climate and soil structure, the drying up of the surface waters from the latest rain event is a process lasting up to a few days, whereas the uptake of water for transpiration from within the soil column can continue for weeks until the *wilting point* is reached. The evolution of the system during and in-between successive rain events is schematically portrayed in Fig. 8.1.

8.1. Isotope change in the land-biosphere-atmosphere interface

To a first approximation, the isotope composition of the input into the terrestrial hydrological systems is often assumed to be invariant with respect to the isotopic composition of the precipitation, thus enabling one to use the GNIP data set as a measure of this input. Such an assumption can be justified in the case of rain in temperate or tropical climates under conditions when there is little water hold-up of the water near the surface and where the incoming flux is rapidly incorporated into a large and well-mixed reservoir, such as groundwater or a large lake. In practice, especially in the more arid as well as in cold climate regions, noticeable changes are on record. Early indications of such changes were reported in the seminal studies on the recharge flux's isotope composition (Halevy, 1970; Gat and Tzur, 1967; Vogel and VanUrk, 1975) as well as on the comparison of the

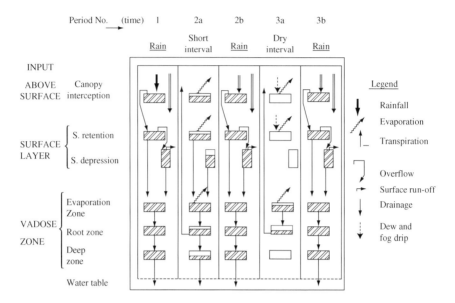

Fig. 8.1. Schematic representation of the response of reservoirs to successive rain events during the transition of precipitation through the atmosphere/plant/lithosphere interface.

surface runoff to the local precipitation under a variety of climate settings (Vogel; Fritz *et al.*, 1976; Levin *et al.*, 1980). This evidence is varied, ranging from more depleted values of the heavy isotope composition in surface runoff following heavy rains in the temperate and arid zones, to enriched values in a semi-arid setting. Enriched δ-values in the groundwater recharge flux are seen in some cases and an opposite trend in others.

Assuming a constant isotopic composition for all incident rainfall, the only change that would be expected during the passage of water through the surface systems is that imposed by partial evaporation from either open water surfaces or from within the soil column and changes due to the melting of snow or ice as discussed in Section 7.3. Since the enrichment of the heavy isotopes during evaporation follows distinct "Evaporation Lines" in δ-space with lower slopes than the "Meteoric Water Lines", the residual waters would then be characterized by lower *d-excess* values than the precipitation.

In actuality, changes in the isotope composition of the precipitation occur both within the time scale of a single shower and from one shower to the next. Moreover, there is usually a pronounced seasonal cycle. Because in most cases, an amount-effect on the isotope composition of rain is observed, the preferential loss of the smaller rains imposes a shift in the isotopic composition of both the runoff and the percolation fluxes relative to the average isotope composition of the precipitation. Such shifts follow the "Local Meteoric Water Line" (LMWL) and may impose a modest change of their own on the *d-excess* value. The degree of such shifts depends on the rain pattern and on the eco-hydrological structure. At times, these "*selection effects*" can overshadow the "*isotope fractionation.*"

The selection between rain events in the case of the groundwater recharge flux on a seasonal basis is especially noticeable due to the larger uptake of soilwater by plants during the warmer summer period so that the cold period rains are preferentially represented in the recharge flux. This effect has been utilised, for example, to characterise the groundwater recharge process in Taiwan based on the very marked seasonal difference between the isotopic composition of the summer and winter rains in the China Sea region (Lee *et al.*, 1999). The selection process is even more marked in the case of solid precipitation or permafrost conditions because of the loss of part of the intercepted snow cover through sublimation or runoff following its melt-down (Sugimoto *et al.*, 2003).

In summary, the change in the isotopic composition of the incoming precipitation during its passage through the land-biosphere-atmosphere interface, the so-called Isotope Transfer Function — ITF, is the result of an interplay between *fractionation* and *selection* effects. Some examples of the changes in isotope content of groundwater recharge and surface runoff relative to the precipitation input under different climate settings are shown in Figs. 8.2(a–d). The first of these (Fig. 8.2a) refers to a temperate zone setting, in which the major control on the occurrence of runoff is the rain intensity which expresses itself as an isotopic signature in the runoff to the extent that there is a marked *amount effect* on the isotopic composition of the precipitation. Fig. 8.2b describes the situation in a semi-arid setting, where surface runoff is only a very minor component of the water balance and, as described by Gat and Tzur (1967), there is an enrichment of the infiltrating flux due to evaporation from the surface. The amount of enrichment of the heavy isotopologues by this process depends on the relative rate of rainfall and infiltration. Fig. 8.2c is an example from the arid Negev Desert where, as will be discussed in more detail later, the deeper percolation into the subsurface is conditional on the accumulation of a sufficient water depth on the surface by means of runoff, and where these surface waters then suffer some evaporative water loss prior to recharge. Figure 8.2d,

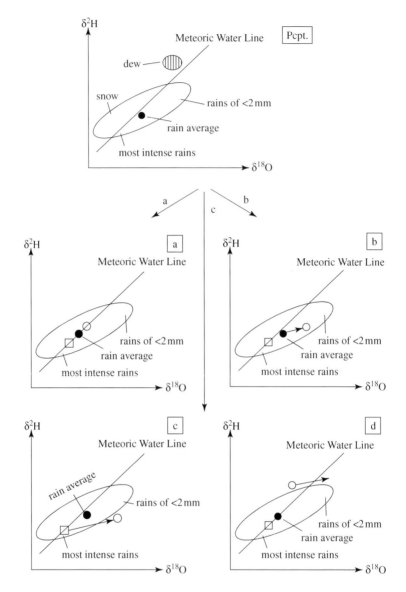

Fig. 8.2. Change of the isotope composition from precipitation to surface runoff and groundwater recharge for different climate settings: (a) temperate zone, (b) semi-arid setting, (c) the arid Negev Desert, (d) region with fog-water input. Open circles represent the local percolation flux, open diamonds represent the average composition of the local surface runoff.

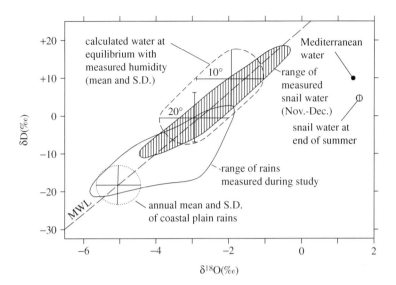

Fig. 8.3. Comparison of $\delta^{18}O$ and δ^2H in snail body-water to environmental waters (based on Goodfriend *et al.*, 1989). The isotopic values for water in equilibrium with the ambient humidity were calculated for 10° and 20°C, respectively.

from a region with abundant fog occurrence shows the secondary effect of recycled evaporated waters with their distinctive d-excessive isotope composition on the land interface water balance. It is of interest to note that a similar case where recycled humidity rather than the direct precipitation input was taken up by snails was also identified through its isotope signature, as shown in Fig. 8.3. These data point to the importance of dew in the water balance of this system.

8.1.1. *Above-surface processes*

On vegetated areas, part of the incoming precipitation is intercepted on the canopy. The holding capacity of the canopy relative to the amount of rain and the rate of evaporation of the intercepted water during the interval between successive rain events, as well as the aerial coverage of the canopy, determine the percentage of precipitation which reach the ground. The evaporative water loss from these canopy-intercepted waters accounts for a sizeable fraction of the water balance in heavily vegetated areas, i.e. up to 35% in the Amazonian rain forest (Molion, 1978) and 20.39% and

14.2% of the total rainfall on coniferous and deciduous canopies, respectively, in the Appalachian Mountains, USA (Kendall, 1993). The waters that eventually reach the ground are made up of the *throughfall*, which consists of the precipitation that finds its way through open spaces devoid of a leaf cover and the excess of water draining from the leaves once their holding capacity is filled up, and the *stemflow*, which is drainage from the canopy that is channeled along the branches onto the stems to the ground.

Another aspect to be taken into account in this respect is the seasonal change in leaf area, notably in the case of deciduous trees and of agricultural crops.

The degree of change of the isotopic composition due to the partial evaporation of the intercepted waters depends on the relative size of the intercepted volume relative to the amount of precipitation and the amount evaporated during the interval between successive rain-spells. It is described by the G-function (see Box 8.1), whose property is such that the maximum effect is seen when about one half of the intercepted moisture is lost to evaporation before the residue is flushed down by the successive rain whereas, obviously, no change is recorded when either none or all of the intercepted moisture is lost.

The case of snow interception is often characterised by an increase in the *d-excess* parameter which is the result of the isotope changes due to the partial snowmelt and sublimation (*vid.* Chapter 7) and thus differs from the isotope signature of the evaporative process which accompanies rain interception.

Even in the absence of precipitation, the vegetation cover can affect the hydrologic balance of the surface and its isotopic signature through the interception of fog and fog drip (e.g. Ingraham and Mathews, 1988), resulting in the slowing down of the drying of the surface soil. Other effects are due to the changes in the surface temperature balance resulting from albedo changes and the effect on the aerodynamic turbulence of the air boundary layer as a function of the surface roughness, which can affect the evaporative signature in open surface waters.

Above-surface interception of precipitation is also encountered in urban built-up areas. Depending on the roof drainage systems, the incoming precipitation is exposed to additional evaporation loss and further often channeled preferentially towards surface runoff. Due to the nature of the intercepting surfaces, this process is accompanied by a likelihood of chemical pollution as well.

Box 8.1. Stable isotope composition of the output from a limited capacity reservoir during sequential phases of wetting and drying.

	Input (precip.)		water in reservoir		output	
	Vol	SIC	Vol	SIC	Vol	SIC
1 — 1st rain event	aV_0	δ_a				
2 — filling-up phase:						
when $a \geq 1$			V_0	δ_a	$(a-1)V_0$	δ_a
when $a \leq 1$			aV_0	δ_a		
3 — partial evaporation (drying up phase): when $a \geq 1$			fV_0	$\delta_f = \delta_a - \varepsilon.\log f$		
4 — 2nd rain event						
when $b \geq (1-f)$:	bV_0	δ_b	V	δ_{Res}	$(b-1+f)V_0$	δ_{Res}

$$\delta_{Res} = (f\delta_f + b\delta_b)/(b+f)$$

The SIC of the output is thus: $[(a-1) \cdot \delta_a + (b-1+f)\delta_{Res}]/(a+b+f-2)$

When $\delta_a = \delta_b = \delta_{Pr}$ the SIC of the water in the reservoir and output (δ_{Res}) can be written as:

$$\delta_{Res} = [(\delta_{Pr} - \varepsilon \cdot \log f)f + b\delta_{Pr}]/(b+f) = \delta_{Pr} - f \cdot \varepsilon \cdot \log f/(b+f)$$

The isotope enrichment during the process is then given by:

$$\delta_{Pr} - \delta_{Res} = \varepsilon \cdot f \cdot \log f/(b+f) = G \cdot \varepsilon$$
$$\text{so that: } G = f \cdot \log f/(b+f)$$

Fig. A. The function $G = \frac{f \cdot \log f}{(b+f)}$ for values of $b > (1-f)$

8.1.2. *Local surface runoff*

Water in excess of the holding or infiltration capacity at the land surface
will discharge as overland flow. This flux as it runs along can be augmented
by additional discharge of interflow and groundwater, or diminished by
evaporation and infiltration along the flow path. The relation between the
isotopic composition of the runoff and that of the precipitation from which
it originates depends on a number of factors: foremost the hold-up in the
surficial reservoirs relative to the variability of the precipitation and, obviously, the change of the isotope composition of the water resulting from
the evaporation. Both these factors depend on the surface structure and
morphology as well as on rain characteristics (intensity, duration and inter-
mittency), respectively.

The isotope signature of the rainfall is transmitted without discrimination to the overland flow, including the large variability in isotope composition during the shower, on exposed rock surfaces as well as on clayey
soils, such as the loess cover of many desert areas and on pavements in built
up areas. Any change along the flow path will then be the result of either
admixture of an additional water source or internal mixing and evaporative
water loss when the water flow is delayed in a surface impoundment, as will
be discussed in more detail in Chapter 9. This situation is most common
in the arid region due to the absence of a continuous plant cover and the
resultant lack of soil accumulation. It is to be noted, however, that in contrast to the situation just described, the rain falling on bare sandy areas
in the arid zone is totally infiltrated, to be later almost entirely lost to
evaporation from within the sand layer. The only surface runoff recorded
in sand-dune areas results when these are covered by algal mats, whose
survival is possible due to the dew deposition on those areas (Yair, 1990).

In the semi-arid climate zone, overland flow is but a minor component of
the water balance and generated only by exceptionally intense rain events.
Due to the "amount effect" on the isotopic composition of the precipitation, this runoff is usually characterised by more depleted δ-values than
the average rain. To a lesser extent, the same is also true for the temperate zone where surface runoff constitutes a large proportion of the precipitation input. Further downstream, the flow is augmented by subsurface
and spring discharges; since these waters are of a mixed character and represent a larger part of the annual rain input, their isotope composition is
closer to that of the annual average, and less variable than the immediate
overland discharge. Indeed, the comparison of the variability of the isotope

composition of the runoff to that of the precipitation is being used to assess the relative contribution of overland flow and interflow on the surface runoff (Sklash *et al.*, 1976). The above will be further elaborated in Chapter 9.

8.1.3. *The local percolation flux*

The local recharge flux consists of the residual waters of a rain event that are left after deducting the surface runoff and the evapo-transpiration losses from both surface waters and from within the soil column. The infiltrating water undergoes a process of mixing with the antecedent moisture content of a whole suit of reservoirs, including those at the surface, within the soil column and perched groundwater bodies. Its isotopic composition reflects a-primo the selection between the different pathways (surface runoff, evapo-transpiration and infiltration, respectively) and is then modified by the isotopic fractionation in accompaniment of any evaporation. Since the percolation and passage through the soil column involves both a delay and mixing with residues of former rain event, the recharge flux unlike the runoff streams is characterized by a rather steady averaged isotopic value, except for the case where by-pass conduits in cracks or sinkholes intervene. The selection on an event or seasonal basis is very pronounced and differs for different precipitation regimes as well as the ecological and land-use pattern concerned.

To a first approximation, the isotopic composition of the recharge flux follows that of the precipitation rather closely in the temperate zones. As measured on *lysimeter* percolates (Halevy, 1970; Sauzay, 1974) or on seepage waters into caves (Eichler, 1965; Harmon, 1979) it appears that the fluctuations of the isotopic composition of individual rain-events are smoothed out by the transition through the surface and soil zone. However, the percolated water are generally observed to be slightly depleted in the heavy isotopes relative to the amount-weighted average of the precipitation, an effect attributed to the amount effect on the isotope content of precipitation since the more intense rains are preferentially represented in the recharge flux, especially so in the by-pass flux through larger conduits. Such an effect was also observed in groundwaters in the semi-arid regions of South Africa (Vogel and vanUrk, 1975). Another feature reported in the literature is the preference of winter over the summer precipitation in the recharge flux resulting apparently from the large water uptake from within the soil column by the vegetation during the growing season, as reported by Lee *et al.* (1999) in Korea. A similar result, but for another reason, is the case of large representation of snowmelt in the annual recharge flux. However, in areas of permafrost in Siberia, as reported by Sugimoto *et al.* (2003), a large part of the snowmelt disappears as runoff as it is prevented from infiltrating the soil due to its frozen condition. Only the spring and summer rains are then used by the plants and as a source for deeper percolation. Indeed, it was shown by the dating of deep and old groundwater that recharge was prevented in ice-covered regions during the last ice age (Beyerle *et al.*, 1998). Evidently, the timing of the precipitation relative to the ecological and hydrological condition needs to be examined in each case.

The changes introduced in the isotope composition of the recharge flux by the soil processes will be further discussed in Section 10.2.1.

Chapter 9

Surface Waters

9.1. The surface runoff

As described in section 8.1.2, the excess water during and following rain events (beyond the infiltration capacity of the soil and the holding capacity of surface depressions) runs off on the surface in a pattern dictated by the surface morphology and ecological structure. The local runoff streams merge downstream to form rivulets that then discharge into the large river systems. As the waters drain over the surface, they interact with the antecedent waters in surface depressions and the topsoil, and are further exposed to water loss by evaporation and, on the other hand, may be augmented by additional precipitation input and the discharge of sub-surface waters such as from *interflow* or groundwater exposures. Both the geochemical and isotopic makeup is fashioned primarily by the precipitation input and its local interactions at the surface. Further modifications can then take place whenever overflow into adjacent wetlands or *bank infiltration* occurs, as well during the sojourn of the river waters in through-flow or dammed lakes.

The relationship between the surface and sub-surface drainage systems is very different in various climate settings, as already described in Chapter 1. In humid tropical climate regions, the direct surface runoff dominates, finding its way through local ponds and wetlands into the regional river system. In contrast, in the temperate climate zones, an appreciable portion of the discharging waters have some history of subsurface drainage, either by interflow from the *vadose* zone, or as groundwater discharge. In the semi-arid zone, the direct surface runoff is minimal and occurs only following a lengthy rain-spell or very strong downpours that exceed the infiltration capacity of the soil surface. Otherwise, the water that is not lost

locally by evapo-transpiration recharges the groundwater and the rivers are
then fed way downstream by larger spring discharges from the aquifers.
Finally, in the arid drylands, the situation is reversed with local surface
flows infiltrating the flow path into subsurface systems so that the surface
streams are reduced rather than augmented as they proceed downstream.

The Isotope Transfer Function (the ITF, defined in section 8.1) from
precipitation to the runoff reflects these different relationships. Generally
speaking, the direct surface runoff is engendered on bare surfaces or
on fully saturated soil when the continuing precipitation is in excess of
the infiltration capacity. In this case, the surface runoff is very often
more depleted in the heavy isotopes than the amount-weighted average of
the rainfall (Ehhalt *et al.*, 1963; Levin *et al.*, 1980). This is attributed to the
"amount effect" on the isotopic composition of precipitation, since only
the heavier and long-lasting rainfall contributes to this direct runoff. The
large variability of the isotope composition within and between showers is
then directly transmitted into the runoff, only later downstream to be attenuated by admixture of more and more waters as well as when the stream
moves through hold-up reservoirs such as ponds, wetlands etc., where the
different inputs are mixed. Any evaporative signature during surface flow
is rather minor in a fast-flowing stream, but may become more significant
during the sojourn in hold-up reservoirs. Fontes and Gonfiantini (1967)
showed, however, that a change in isotopic composition was recorded even
for evaporation from a flowing stream under arid conditions in the deserts
of North Africa.

It was shown by Sklash *et al.* (1979) that the runoff in the temperate zone may involve a phase of interflow through the soil column and
therefore both a delay and mixing of the precipitation pulses. It is thus less
variable and its isotope composition closer to that of the rain average.
Since there is, in this case, also less of an exposure to evaporation on the
surface, there is little sign of an evaporative signature, except so far as it is
inherited from the above–surface processes on the canopy or due to a pre-
event effect on the antecedent surface or soil moisture. Obviously, in detail,
the local precipitation characteristics as well as the surface structure and
morphology, determine the final outcome.

9.2. Isotopic hydrograph separation

The varying flow in the runoff stream on either an event or seasonal basis is
the response of both the direct surface drainage and spring discharge (the

base flow) to the precipitation input. The basic premise for the use of the isotopic tracers in identifying these two components in the runoff stream, i.e. the *hydrograph separation*, is that the base flow consists of groundwater discharge whose isotope composition is steady and represents an averaged value of the groundwater recharge over the recharge zone, whereas the episodic direct runoff is conditioned by the relevant rain event with a possible modification by the flushing of accumulated surface and soil water residues. It should be noted, however, as discussed above, that in the temperate zone, even the relatively direct runoff occurs by means of *interflow* and thus its isotope composition also represents an averaged value of some antecedent rain events.

During the period that the atmospheric Tritium values changed appreciably from year to year (*vid.* Appendix b), the Tritium value in the runoff compared to that of the recent rainfall and the groundwater bodies was used to great advantage for such a purpose, as exemplified in the study of the sources of the Jordan River by Simpson and Carmi (1983).

The contribution of snowmelt from seasonal snowpacks can also be identified by the distinctive isotopic signature of very depleted (negative) isotopic values often accompanied by a slight evaporative signature.

9.3. River systems and the "Isotopic River Continuum Model"

The isotopic composition of an undisturbed flowing river is the amount-weighted average composition of its surface and sub-surface inflows and tributaries. This usually has a pronounced temporal and seasonal pattern reflecting the hydrology of the system: the more steady base-flow is fed as a rule by the subsurface discharge and contributes an averaged isotope signature of its source region, whereas the more changeable flood-flow portion carries in part the signature of a single shower. Indeed, the characteristic different patterns of the isotopic signature of the surface and subsurface discharges as fashioned by the eco-hydrological characteristics of the source areas (as described above) is being used as a tool in the hydrograph-separation of the flow regime in the tributaries (Kendall and McDonnell, 1993), together with their geo-chemical markers. Further, as more and more tributaries discharge into the river as it proceeds downstream from the headwaters, the changing isotope composition can be a measure of their relative contribution to the flow, provided of course that the isotopic composition of each such tributary is distinguishable from the rest and that the river sampling is performed far away downstream from any confluence (the mixing length) to allow for homogeneity across the cross-section of the river. Once this is the case, one can sometimes detect the diffusive seepage of groundwater or of bank-storage discharge into the river by a discontinuity of the isotope composition along the river bottom or banks.

In addition to the identification of the natural or anthropogenic origin of the confluents, the most immediate use made of the monitoring of the evolution of the isotope composition along a river is the quantification of any water loss by evaporation. However, along the course of the river there occur a number of processes which change the simple pattern presented. Among them are:

a)– the holdup and mixing in throughflow lakes, dams or reservoirs,
b)– the diversion of water into adjacent wetlands or bank storage during high water stand and their subsequent release back into the river,
c)– water loss by infiltration into the bottom sediments or by pumping to users outside the watershed,
d)– water extraction for the purpose of irrigation and urban/industrial usage within the watershed, to be returned in part from irrigation return flow or urban effluent discharge,
e)– the import of waters from outside the watershed, either directly into the river system or as return flows from local consumers.

As far as their effect on the isotopic labels of the river flow, one can distinguish between processes that result in mixing and averaging of the temporal variability (processes a, b and d) either by ponding or by extraction and delayed return to the river, others that change the natural water balance by irreversible loss of water or by addition of extraneous water sources (processes c and e) and, finally, those processes that result in change of the isotope composition by exchange with stagnant water, such as during bank storage, and changes resulting from the fractionation between the water *isotopologuess* during evaporation (processes a, b and d). The latter process, namely the enrichment of the heavy isotopic species in a water body exposed to evaporation and especially, the resultant decrease in the value of the *d-excess* parameter is being routinely used to quantify the water balance in such systems. While this is a rather straightforward procedure whenever the evaporating waters and the ambient moisture are in isotopic equilibrium one with the other (as is the case generally in the headwater region), this, however, is not true anymore along the river course; there, the isotope composition of the inflow flux into the evaporation sites is inherited from the upstream part of the river system and mostly not in equilibrium with the local atmosphere.

In analogy to the "River Continuum Concept" (Vanotte *et al.*, 1980) the *"Isotope River Continuum Model"* (IRCM) was proposed with the aim of identifying the isotopic parameters which can be employed to characterise the water balance and river/environment interactions along the river course. The scheme of the IRCM is shown in Fig. 9.1. The change in the *d-excess* parameter along the flow path, rather than the change in the

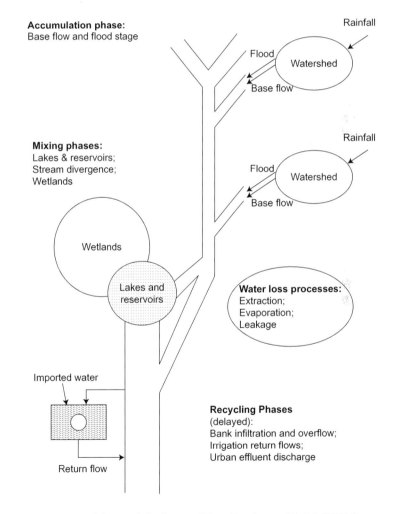

Fig. 9.1.   Scheme of the Isotope River Continuum Model (IRCM).

δ-values per-se, yields a measure of the amount of the evaporated water. The non-evaporative water loss by pumping or other extraction of water from the river is not directly recorded in the isotopic makeup, but manifests itself through a larger shift in the isotope composition along the evaporation line for a given amount of evaporation. For a total water balance of the system, therefore, both discharge rates and the isotope composition have to be monitored along the watercourse.

A detailed study of the tropical Piracicaba River basin by Martinelli *et al.* (2004) exemplifies some of these relationships. It was found that the isotope content of the river flow during the rainy season responds to flood-flow events, showing a notable variability around the averaged yearly value which is better conserved during the drier summer, when the flow is essentially one of baseflow from the discharge of the groundwater aquifers. Notable local changes are the result of discharges from anthropogenic sources; the change of the *d-excess* value is then more easily identified than the δ-values, which are more variable even under undisturbed natural conditions.

The upper Jordan River, as reported by Gat and Dansgaard (1972), shows the sensitivity of the system's response to evaporative water loss on the detailed flow scheme of the river through the maze of the Hula Swamp area, the reason being the dependence of the evaporative enrichment in the heavy isotopes on the ratio of the evaporation flux to the volume of the evaporating system.

9.4. Open waters (ponds to lakes)[*]

The enrichment of the heavy isotopes in the residues of evaporating water bodies imparts a characteristic signature on these waters, which enables the tracing of their movement within the hydrologic cycle, in addition to serving as a tool for estimating the water balance of such systems. The isotopic balance of a well-mixed open water body of volume V_L, whose material balance is given by Eq. (9.1), is described in Eq. (9.2); F(in), F(out) and E are the amounts of inflow, outflow and evaporation, respectively.

$$dV_L/dt = F(in) - F(out) - E \qquad (9.1)$$

$$d(V_L \cdot \delta_L)/dt = F(in) \cdot \delta_{in} - F(out) \cdot \delta_L - E \cdot \delta_E \qquad (9.2)$$

---

[*]The isotope enrichment in lakes has been extensively reviewed over the years, the more recent ones being Froehlich, Gonfiantini and Rozansky (2005); Rozanski *et al.* (2001) and Gat (1995).

This formulation implies that only the evaporation process is associated with an isotope fractionation and that the water discharging from the lake, whether by liquid outflow or percolation and leakage into the bottom or lakeside banks, does not involve any discrimination between the isotopic species. The incoming flux can consist of a river inflow and direct runoff, groundwater discharge or additional precipitation on the liquid surface, and a weighted average of the isotope composition of the different fluxes obviously has to be introduced into Eq. (9.2). Based on the value for δ_E as given in Eq. (4.9a), the buildup of the isotopic composition in a lake at hydrologic steady state (*i.e.* constant V_L) can then be written as follows:

$$d\delta_L/dt \approx [F(in)/V_L] \cdot [\delta_{in} - \delta_L] - [E/V_L] \cdot [\{h(\delta_L - \delta_a) - \varepsilon\}/(1 - h)] \quad (9.3)$$

δ_a refers to the isotopic composition of the atmospheric moisture, and h the humidity in the overlying atmosphere relative to that of the saturated vapour for the liquid at its surface temperature and salinity. The equation applies to either the hydrogen or oxygen isotopes, provided the appropriate parameters are introduced into this equation.

One can consider 3 types of systems:

- a throughflow system, with both inflow and outflow; in such a system a hydraulic steady state and constant volume is achieved when the inflow is balanced by the sum of the evaporation and outflow. The isotope composition of the evaporating system will build up as evaporation proceeds until a steady isotope composition is reached. The degree of buildup ($\Delta = \delta_{ss} - \delta_{in}$) as given in Table 9.1 is a function of the humidity and the hydrologic balance, *i.e.* the ratio of $F(in)/E$, as shown in Fig. 9.2.

- a terminal system, namely a system without liquid outflow, which reaches steady state when the inflow is balanced by the evaporation flux. This system can also reach an isotopic steady state once $\delta_E = \delta_{in}$. It has to be considered, however, that such systems usually accumulate the introduced salts since they are not flushed out by the evaporation flux. Because, as discussed in Chapter 11, the evaporation flux and isotopic buildup both change with the increasing salinity, the steady state situation is usually but a transient one.

- an isolated desiccating pond. In this case no hydrologic steady state exists, but as the isotope content builds up in the desiccating system an isotopic steady state is reached when $\delta_E = \delta_L$.

Isotope Hydrology

Table 9.1.   The enrichment of the heavy isotopes in lake systems under steady state conditions.

A through-flow lake at hydrological and isotopic steady state:

$$\Delta\delta = \delta_{ss} - \delta_{in} = [(\delta a - \delta_{in} + \varepsilon^*/h + (1-h)\cdot\theta\cdot n\cdot C_k/h]/[1 + \{F_{(in)}/E\}\cdot\{(1-h)/h\}]$$

when: $(\delta a - \delta_{in}) = -\varepsilon*$, then:

$$\Delta\delta = (1-h)\cdot(\varepsilon^* + \theta\cdot n\cdot C_k)/[h + (1-h)\cdot(F_{(in)}/E)]$$

A terminal lake at hydrological and isotopic steady state:

$$\Delta\delta = \delta_{ss} - \delta_{in} = h\cdot(\delta a - \delta_{in}) + \varepsilon^* + (1-h)\cdot\theta\cdot n\cdot C_k$$

when: $(\delta a - \delta_{in}) = -\varepsilon^*$, then: $\Delta\delta = (1-h)\cdot(\varepsilon^* + \theta\cdot n\cdot C_k)$

A desiccating pond at isotopic steady state:

There is no hydrologic steady state in this case

$$\delta ss = \delta a + \varepsilon^*/h + (1-h)\cdot\theta\cdot n\cdot C_k/h$$

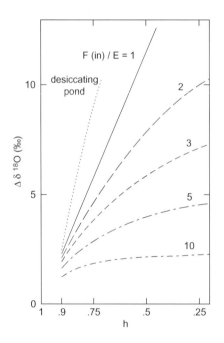

Fig. 9.2.   The isotope buildup in through-flow lakes at hydraulic steady state as a function of humidity and the ratio F_{in}/E. Isotopic equilibrium between the inflow waters and atmospheric moisture is assumed. The buildup for a desiccating pond (e.g. an evaporation pan) is shown for comparison.

The buildup of the isotope composition in these three systems at steady state is given in Table 9.1.

Comparing the relative enrichment of the isotopologues of Deuterium and Oxygen-18, respectively, in the evaporating waters, namely the slopes of the Evaporation Lines in δ-space, i.e. $S_{EL} = \Delta^2\delta/\Delta^{18}\delta$, yields the following relationship:

$$S_{EL} = \frac{[h \cdot (\delta_a - \delta_{in}) + \varepsilon^* + (1 - h) \cdot \theta \cdot n \cdot C_k]_2}{[h \cdot (\delta_a - \delta_{in}) + \varepsilon^* + (1 - h) \cdot \theta \cdot n \cdot C_k]_{18}} \qquad (9.4)$$

The slope is a function of the relevant isotopic fractionation factors ε^* and C_k, and as shown in Fig. 9.3 as well of the humidity and of $(\delta_a - \delta_{in})$. However when the atmospheric moisture and the input waters are in isotopic equilibrium one with another, *i.e.* when $(\delta_a - \delta_{in}) = -\varepsilon^*$, then the slope of the Evaporation Line is given by the relationship:

$$*S_{EL} = [\varepsilon^* + \theta \cdot n \cdot C_k]_2/[\varepsilon^* + \theta \cdot n \cdot C_k]_{18}. \qquad (9.4a)$$

This slope is no longer a function of humidity and because the C_k values of both the deuterated and Oxygen-18-labeled water molecules are rather similar, the slope of the Evaporation lines is much smaller than that of

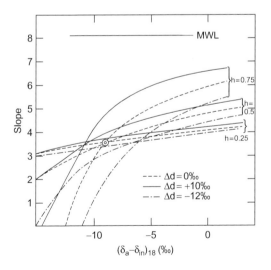

Fig. 9.3. The slope $(\Delta\delta^2\mathrm{H}/\Delta\delta^{18}\mathrm{O})$ of the Evaporation Lines from an open water surface as a function of ambient humidity and the difference in isotope composition of the inflow and the atmospheric moisture, for three cases of $\Delta\mathrm{d} = (\mathrm{d}_a - \mathrm{d}_{in})$, as indicated ⊚ signifies the condition of Equation 9.4a.

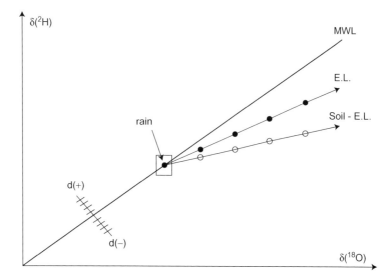

Fig. 9.4.   Evaporation Lines for the case of evaporation from a free water surface (EL)
and a soil column (SEL), respectively. MWL signifies the equilibrium line with a slope
of S = 8.

the MWL whose slope in δ-space is conditioned by the ratio of the thermodynamic (equilibrium) fractionation factors, i.e. $S_{MWL} = \varepsilon^*_2/\varepsilon^*_{18}$, and
thus close to the value of 8. As was discussed in section 4.3, the value of
n depends on the aero-dynamic conditions over the evaporating surface, so
that the S_{EL} values given by Eq. (9.4) differ for example for the case of evaporation of an open water body into a fully turbulent atmosphere from that
of evaporation under windstill conditions and even more so into a stagnant
air layer such as within the soil, as shown in Fig. 9.4. In Eqs. (9.4) and
(9.4a), it is usually assumed that $\theta \approx 1$; this is not necessarily true over
large evaporation systems, as will be discussed below.

The formulations given above require that a steady averaged value of the
isotopic composition of the inflowing waters and the ambient atmosphere
can be applied. However, in many cases, the parameters that govern the
isotope composition of the evaporating waters undergo changes with time,
seasonal or even inter-annual ones. The evaporation rates peak during the
warm period and the isotopic composition of precipitation goes through an
annual cycle. The latter affects that part of the term δ_{in} that results from
direct precipitation input and surface runoff, while the groundwater discharge yields a more averaged and less variable input. As is evident from

Eq. (9.3), the response of the system's isotope composition to changing input parameters depends on the size of the reservoir, more precisely on the residence time of water in the system, which can be estimated by the term $V_L/F(in)$. Thus, in a shallow lake, the water will at any time be close to the steady isotope composition for the momentary set of parameters, and the isotope composition of the lake waters will vary in an extreme manner as these parameters change. On the other hand, the larger the mixing volume, the more restrained the response for any given environmental change. Indeed, for lakes whose residence time is of the order of years, the composition of the lake just fluctuates around the mean annual isotope composition, but may drift in response to inter-annual changes in the lake's water balance, as exemplified in Fig. 9.5 for the case of Lake Tiberias, where the mean residence time of the waters is of the order of 5 years (Gat, 1970).

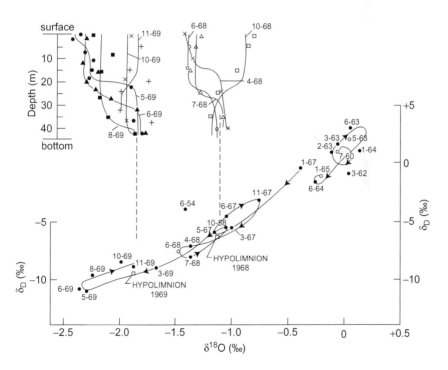

Fig. 9.5a.   Evolution of the isotope composition of the surface waters in Lake Tiberias, showing seasonal cycles and variations during successive years with different water balances. The vertical $\delta^{18}O$ profiles during the years 1968 and 1969 are shown on top.

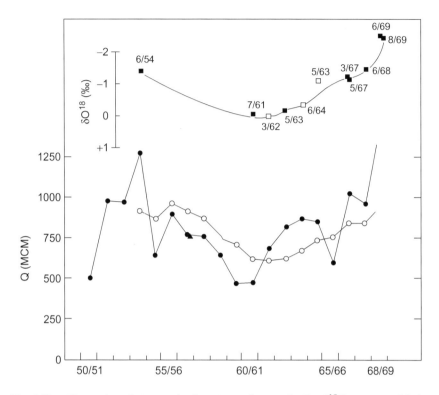

Fig. 9.5b.   Comparison between the long-term changes in the δ^{18}O content of Lake Tiberias waters (upper curve) and the 5-year average of the amounts of annual inflow (open circles in lower curve); Full circles show the annual inflow during preceeding hydrological year, i.e. October to September.

The relationships discussed above apply strictly to systems where there is no divergence of the vertical flux over the evaporating surface, implying that the atmospheric domain does not change measurably all over the evaporating surface water. In reality, over larger water bodies, the moisture content and its isotopic composition are modified by the incorporation of the evaporation flux into the overlying atmosphere, as shown in the scheme "Lake Model" of Fig. 5.7. In terms of the equations describing the buildup of the enrichment of the heavy isotopes in the surface waters, this feature is expressed by a reduction of the value of the parameter θ in the expression for $\Delta\varepsilon = (1 - h) \cdot \theta \cdot n \cdot C_K$. As an example, a value of $\theta = 0.88$ was estimated as the average in the case of the evaporation from the Great Lakes of northern America during summer (Gat *et al.*, 1994).

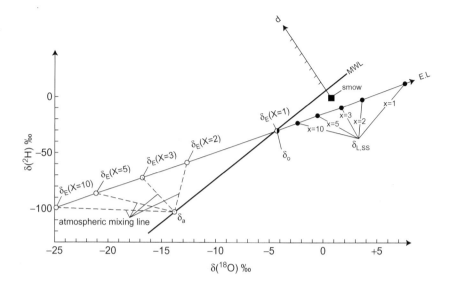

Fig. 9.6. The isotopic composition of atmospheric moisture over an evaporating water body of initial isotope composition δ_o, exemplified by lakes with different water balances (as expressed by the ratio X of inflow to evaporation) and buildup of the heavy isotope content. $\delta_{L,SS}$ signifies the respective steady-state isotopic content for the different value of X. δ_E is the corresponding isotope content of the evaporate which then mixes with the ambient moisture (δ_a) along the indicated mixing lines.

As the evaporate mixes into the ambient air, it modifies the isotopic composition of the atmospheric moisture, especially its d-excess value. As shown in Fig. 9.6 the atmospheric mixing lines depend on the degree of evaporative buildup in the evaporating water body. The change in the d-excess in the atmospheric waters is further discussed in Box 13.2.

9.4.1. *The Index-Lake concept*

Of all the terms in the isotope balance equation, Eq. (9.2), the value of δ_E cannot be measured directly. It may then be estimated on the basis of the Craig-Gordon Evaporation Model using simplifying assumptions that cannot usually be exactly verified. One way to overcome this quandary is to perform isotopic measurements on a water body whose water balance is fully controlled and known, and which is exposed to identical environmental conditions as that of the studied system. The classical tool in this respect is an *Evaporation Pan* that is situated on the edge of the lake or on a platform or island within the lake, such as is used by hydrologist to measure the

"Potential Evaporation" rate. Based on the enrichment of the isotopes in the pan relative to the feed water, it is then possible to estimate the value of δ_E for conditions close to those applying to the system under investigation. This approach has been relatively successfully applied by Allison and Leaney (1982) and Gibson *et al.* (1999) using constant–volume Class A evaporation pans, by following the response of the isotopic composition of the water in the pan to the changing environmental conditions. Because of the small size of the system, it follows rather closely the steady-state concentration for the environmental conditions at any time. Recently, Froehlich (2000) used the rate of achieving the steady state in a desiccating pan for a similar purpose.

It is well known that the evaporation from pans overestimates the evaporation rate of the natural systems, due mainly to the increased turbulence resulting from the edge effect of the walls of the pan and further due to a different temperature regime in the free-standing pan. The latter has been improved by using pans that are thermostatted by a jacket through which the lake water is circulated (Gat, 1970).

An alternative approach, which was first suggested by Dincer (1968) for use in areas with a number of lakes, is the concept of an "index lake." One chooses one of these lakes, preferably a terminal system whose water balance is unequivocal and whose isotope buildup at steady state is characterized only by the relevant atmospheric variables and not by the hydrological ones (*vid.* Table 9.1), in order to obtain the relevant parameters for estimating the value of δ_E including the pertinent value of δ_a. Gat and Levy (1978) in a study of coastal salt pans in the Sinai Desert region used the steady state isotopic buildup in a drying-out pond (without inflow or outflow) for the same purpose. The isotopic steady-state composition of such a desiccating system, which is given by the relationship:

$$\delta ss = \delta a + \varepsilon^*/h + (1 - h) \cdot \theta \cdot n \cdot C_k/h,$$

is fixed only by the atmospheric variables and the relevant fractionation factors and thus is a very suitable choice.

9.4.2. *Vertical mixing in lakes*

In reality, the isotope composition of most lakes is not really homogenous. The incoming stream can often be traced for considerable distances before being incorporated by mixing in the water body; in particular, in the case of a density difference between the inflow and the surface waters due to the temperature or salinity. The horizontal mixing by means of surface currents and waves is rather rapid but the vertical homogenization in the case of stratification between the surface layer (the epi-limnion) and the deeper layers (the hypo-limnion) can be delayed for long periods of time; for a detailed discussion *vid.* Imboden and Wuest (1995). Typically due to the heating of the surface layer, a seasonal stratification is established during summer, later dissipated in autumn as the temperature gradient is again reduced (the mono-mictic regime). Since the isotopic composition of the surface layer undergoes a yearly cycle actuated by the seasonally changing influx and evaporation patterns, the vertical isotopic profile is a valuable monitor of the vertical structure and the mixing efficiency. An example for such a seasonal cycle is shown in Fig. 9.5a. An interesting special case is that of ice covered lakes, where the isotopic signature in the water profile in winter results from the isotope fractionation during the formation of ice, which is conserved in the liquid column since the ice cover shields the liquid from wind-induced mixing. This effect and the isotopic make-up of the ice cover itself is described below in section 9.4.3.

In deep lakes and especially in saline lakes where there is a large density difference between the freshwater inflow and the deeper (more saline) water layers, there develops a long-term stratification (meromixis) which may last for centuries. As shown in extensive studies of the Dead Sea, the deepest terminal lake, the upper layer (the *epilimnion*) undergoes an annual cycle which is not only determined by the balance between the freshwater influx and its evaporative enrichment but even more so the depth of the mixed layer. This is minimal during the time that flooding occurs in winter and thickens towards the end of summer when the density gradient is mitigated.

Due to a reduced freshwater supply to the Dead Sea since the second half of the last century its surface level has been receding continuously, accompanied by an increase in the salinity of its surface waters. This process first resulted in a lowering of the isotope enrichment of the surface waters throughout the seasonal cycle due to the salinity effect on the apparent humidity (*vid.* Chapter 11). Subsequently, as the salinity gradient between the deep layer and the epilimnion diminished, the latter gradually increased in depth during the summer months until finally complete mixing of the entire water column occurred (Steinhorn *et al.*, 1979). These processes have been reviewed in the monograph entitled "The Dead Sea: the Lake and its Setting" (Niemi, Ben-Avraham and Gat, edtrs), published in the series 'Oxford Monograph on Geology and Geophysics. No. 36, and by Gat (1974).

9.4.3. *Ice covered lakes*

Lake ice is a dominant feature of high latitude and alpine lakes over much of the winter season. Supposing that the ice is formed by freezing of the surface water under equilibrium conditions, then the isotope composition of the ice is expected to be enriched by about 3.6‰ in $\delta^{18}O$ along a line of slope $S = \Delta^2\delta/\Delta^{18}\delta = 5.9$ in δ-*space*, based on the fractionation factors as reported by Majoube (1971). As a result, the residual waters are depleted in the heavy isotopes, in opposition to the effect of the enrichment resulting from evaporation, as is indeed clearly reflected in the difference between the profile of isotopes in the waters of Sparkling Lake, Wisconsin, during summer and winter, respectively (Fig. 9.7, based on Krabbenhoft *et al.*, 1970).

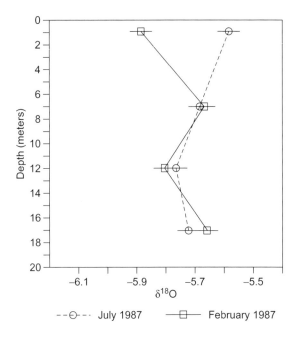

Fig. 9.7.   Vertical profiles of $\delta^{18}O$ in an ice-covered lake in winter compared to the summer profile.

It was to be expected that as the ice cover thickens incrementally during the winter, it could serve as a cumulative record of the changing composition of the lake water, thus enabling us to see whether the through-flow of groundwater persists during the winter months. An examination of the isotopic composition of the ice cover on one of the Wisconsin lakes (Lake Fallison, reported by Bowser and Gat, 1995) showed a complex situation. The 50 cm-thick ice cover had a banded structure. The isotope content showed a marked discontinuity about halfway through the ice cover: in the bottom part, the isotopic values are in between that of the lake waters and the value to be expected for ice, which is formed in equilibrium with the lake water; in the upper part of the profile (the top 25 centimeters), the isotopic composition is that of a mixture of the snow accumulation on top and the lake waters. These findings are presented in Box 9.1.

Box 9.1. Ice cover on Lake Fallison, Wisconsin.

The winter ice cover on Lake Fallison in northern Wisconsin was sampled in mid-winter (March 1989) by coring, as reported by Bowser and Gat (1995). At that time the ice cover was about 50 cm thick with an additional snow-blanket of 10 cm on top. The ice core was subdivided into 12 sections of about 5 cm each and then analysed for their stable isotope and salt content. At that time, the lake water under the ice cover was relatively enriched in the stable isotopes, indicating an evaporative history before the winter:

$$\delta(^{18}O) = -4.06\%_0; \quad \delta(^{2}H) = -44.23\%_0; \quad d = -11.7\%_0$$

Assuming formation of the ice under equilibrium conditions, the isotopic composition of the ice would then be expected to be further enriched in the heavy isotopic species (Majoube, 1971) being:

$$\delta(^{18}O) = -0.5\%_0; \quad \delta(^{2}H) = -23.05\%_0; \quad d = -19\%_0.$$

The isotopic composition of the unconsolidated snow cover, as expected, was very depleted with a mean value of:

$$\delta(^{18}O) = -21.73\%_0; \quad \delta(^{2}H) = -152.63\%_0; \quad d = 21\%_0.$$

As can be clearly seen in Diagram A below, the isotopic composition of the top 24 cm ranged from $\delta(^{18}O)$ values of $-8.9\%_0$ to $-11.5\%_0$ ($\delta(^{2}H)$ values of -67.5 and $-84.5\%_0$, respectively) and the data are situated on a mixing line between the lakewater and the snow composition, as can be observed on Diagram B. The ice in the bottom part of the ice core was more enriched ($\delta(^{18}O)$ values of $-3.65\%_0$ to $-1.5\%_0$, $\delta(^{2}H)$ between -34.3 and $-27.2\%_0$) and spread out between the isotope composition of the lake water and of the estimated value of ice at equilibrium with the lake water.

(Continued)

Box 9.1. (*Continued*)

Based on the end-values given above, the percentage of unfractionated lake water in the lower part of the core ranged from 20% to 50%, and reached up to 60% in the upper part of the core. These values are appreciable higher than those estimated on the basis of the chloride content in the ice (which were close to only 20% in the lower part of the core) based on the assumption that ice formed by freezing under equilibrium conditions excluded all salts. It has to be recognised, moreover, that since no continuous monitoring of the snowfall throughout the season was available, these data are indicative of the overall pattern but a quantitative verisimilitude is not to be expected.

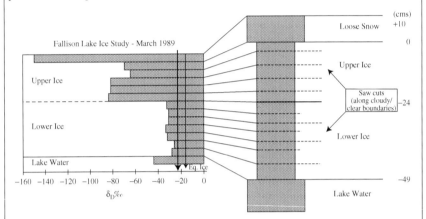

Fig. A.   The Deuterium content of the ice core, as a function of depth. Also indicated are the values for the snow cover and the non-frozen lake water.

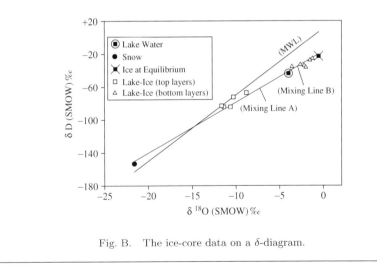

Fig. B.   The ice-core data on a δ-diagram.

Evidently, the lower and upper parts of the ice cover are the result of quite different processes. The lower part appears to be ice grown in equilibrium with the freezing lake water which is mixed with unfractionated lake water that is apparently trapped in the ice during the freezing process. In contrast, the upper part of the ice core is apparently formed by lake water which overflows on top of the subsiding ice bed and mixes with the accumulated snow on top of the frozen lake surface before being refrozen without any further isotope fractionation. As a result of these complex processes which include the incorporation of unfractionated lake water, the ice also contains some of the salinity in these waters that would be excluded by the slow, equilibrium freezing process; the remanent salinity can thus quantify the degree of this admixture. The detailed process obviously depends on the particular climate pattern, in particular the heating and cooling cycle as well as amount of accumulated snow during that time.

Similar measurements on sea-ice indicated that the ice there is derived from the incorporation of deposited snow into freezing seawater (Redfield and Friedman, 1969).

9.4.4. *Lakes without surface outflow*

In areas with a high water table, the groundwater can be exposed in local depression areas, forming a lake without surface outflows — so-called closed lakes. If there is also no surface inflow, such lakes are termed "groundwater lakes." Since the in- and outflow fluxes cannot then be gauged directly, the isotope balance is then often the best method for quantifying the water balance of such lakes. In principle, the increase in salinity due to the loss of water in the lake could also be used for such a purpose; however in many cases the salinity of the groundwater influx into such a lake can be defined less reliably than that of the isotope composition of the groundwater influx.

The comparison of the salinity and isotopic buildup in lakes without surface outflows is an important tool in distinguishing between truly terminal systems (where the only path of water loss is by evaporation) and leaky systems in which there is a hidden subsurface outflow. In this manner, it was found that in the case of the African Lake Chad, there is an overflow into an adjacent evaporative basin (Fontes and Gonfiantini, 1967) as is the case in the Caspian Sea from which waters percolate into the Kara Bogaz basin. It has been shown, however, by Langbein (1961), that an alternative mechanism could be operative in preventing the salinity buildup in a closed lake, namely the formation of spray droplets which are then deposited outside the lake boundaries.

When the evaporative signature of the exposed waters is then transmitted into the down-stream aquifer, it forms an easily identifiable plume of
isotopically enriched waters. The study of Sparkling Lake in north-western
Wisconsin by Krabbenhoft *et al.* (1994) can serve as a good example. The
"evaporative signature" can also often be identified further downstream as
a component of the spring discharges from these aquifers.

9.5. Coupled and complex evaporative systems

As shown above, as long as the isotope composition of the water influx
and that of the ambient moisture can be clearly specified, it is a relatively
straightforward procedure to relate the enrichment of the heavy isotopologues in a lake to its hydrological attributes. This is especially so when the
isotopic composition of the influx is equivalent to that of the local precipitation. In many actual cases, however, the evaporating systems are coupled
and interconnected with one system affecting either or both the influx and
the atmospheric moisture of the adjacent systems. Six such flow schemes of
coupled evaporation elements with different feedback loops were described
by Gat and Bowser (1991) as shown in Fig. 9.8. It is to be noted that
while the rate of buildup of the evaporative signature along the evaporation
elements in a series of element is a function of the hydrologic parameters
F_{in}/E, the final enrichment is a function of humidity but does not reflect
any more the details of the hydrological flow scheme (Fig. 9.9).

The water flux from a complex evaporative system such as wetlands or
swampy terrains, can consist of at least 4 components, namely:

• surface outflow
• sub-surface outflow by leakage into the bottom or banks
• evaporation
• transpiration.

The first two transport processes incorporate both the water and salinity
found in the open water without any fractionation, thus reducing the
amount but not the concentration of the isotopic or geochemical tracer
(with the possible exception of the adsorption and accumulation of some
chemicals on the solid interface, a fact utilized to detect the location of such
leakages). Evaporation results in the increase of the salinity in the liquid
residue accompanied by the enrichment of the heavy isotopes, whereas
transpiration by either aquatic plants or vegetation growing at the water
banks does not fractionate between the isotopic species (*vid.* Section 8.3)

Model A

Feedback to atmosphere:
constant input

Model B1

Feedback to atmosphere:
equilibrium input at each
stage

Model B2

With Rayleigh rainout

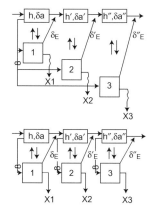

Model C

String-of-lakes, no feedback
to atmosphere

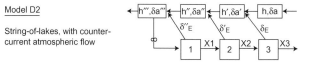

Model D1

String-of-lakes, with feedback
to atmosphere

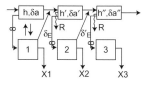

Model D2

String-of-lakes, with counter-
current atmospheric flow

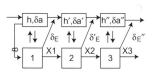

Fig. 9.8.   Flow schemes for coupled evaporation elements with varied feedback loops. In each scheme the upper series of boxes represent the atmospheric reservoirs and the lower boxes represent the evaporating elements.

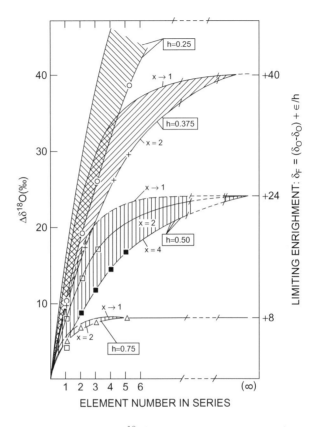

Fig. 9.9. The isotope enrichment ($\Delta\delta^{18}O$) along an evaporative series (for string-of-lakes Model C) for different values of humidity and $X = F_{in}/E$. Open symbols — for $X = 2$; filled symbols — for $X = 4$; the values for a terminal lakes system, $X = 1$, are shown by stippled lines. The ultimate enrichment obtainable is shown on the right-hand side.

but leaves the salinity to accumulate in the liquid residue. Measuring the change of both the isotopologues and salinity enables a distinction between these processes, as was used in an elegant study in the swamps of Okavonga in Botswana by Dincer *et al.* (1979).

Unlike the increase in salinity, which is in direct proportion to the cumulative water loss by evapo-transpiration, as long as the solubility limit of the salts is not reached, the heavy isotope buildup is actuated only by the evaporative component of the evapo-transpiration flux and its degree depends on the antecedent buildup of the heavy isotopes in the system in relation to the atmospheric moisture's isotopic composition. As a result the detailed water pathway and balance need to be considered, as was shown by Gat and Matsui (1991) in a study of the *varzea* in the Amazonas. All this is discussed in more detail in Box 9.2.

Box 9.2. The isotope signature in runoff from a terrain where both evaporation and transpiration occurs.

The water efflux from surface waters, namely by either surface and subsurface discharge or by evapo-transpiration, imparts in part a distinctive signature to the remnant waters.

While both surface runoff and bottom leakages carry with them the isotopic and chemical composition of the water essentially unchanged, both evaporation losses and transpiration result in the accumulation of the ionic constituents in the remaining water. However, direct evaporation results in enrichment of the heavy isotopes of both hydrogen and oxygen (along Evaporation Lines) in the remnant water, while transpiration water usage does not, to a first approximation, fractionate between the isotopic species.

Whereas the salt accumulation due to the combined Evapo-Transpiration flux is proportional to the fraction of water lost, the buildup of the isotopic signature due to evaporation depends on the relative fraction of the water lost, as described in Chapter 4. This property was utilised in the seminal work of Dincer, Hutton and Kupee (1979) to determine the proportion of evaporative and transpired water loss from the Okavango Swamps in Botswana by comparing the buildup of salinity and Oxygen-18 in the runoff compared to the inflow waters. Figure A shows the buildup of the heavy Oxygen isotope as a function of the salinity increase (expressed as the ratio of Electrical Conductivity, EC, in the outflow to the initial value, EC_0) for the different ratios of the Evaporation to Transpiration fluxes. It is assumed that these two fluxes operate simultaneously on a mixed water volume. Two opposing effects can be noted, namely the fact that because of the water loss due to the transpiration, the evaporative enrichment effect is distributed over a decreasing volume, thus enhancing the effect, and the opposite one that the limit of enrichment is reached sooner.

Figure B shows three alternative flow schemes in a runoff basin in which the same amount of the E and T fluxes operate in different sequential order.

Figure C shows the different buildup of the isotope signature under these different scenarios (based on Gat and Matsui).

(Continued)

Box 9.2. (*Continued*)

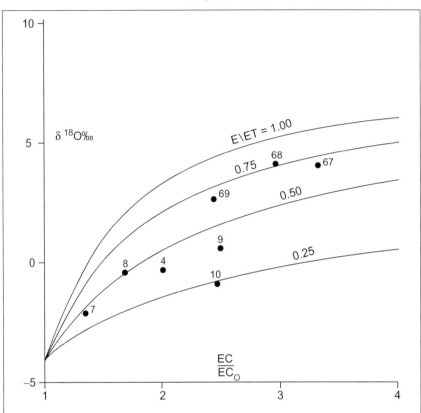

Fig. A.   The buildup of $\delta^{18}O$ in the Okavango Swamps as a function of the salinity increase, for different ratios of the evaporation to the total evapo-transpiration flux. The points shown are data measured by Dincer *et al.* during the summer.

(*Continued*)

Box 9.2. (*Continued*)

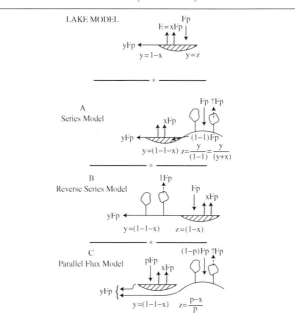

Fig. B.   Flow schemes of runoff through a wetland terrain in the Amazon Basin, consisting of plant covered land and open water areas. F_p — the rainflux; y — the total runoff fraction from the terrain; x and t — the fraction recycled by evaporation and transpiration, respectively.

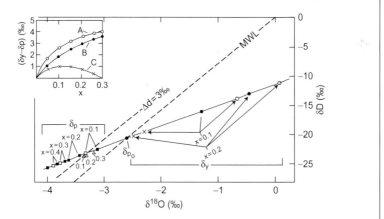

Fig. C.   The isotopic composition of precipitation (δ_p) and runoff (δ_y) in the different E-T schemes shown in Fig. B. assuming a runoff fraction of 0.5 for the basin and an ambient humidity of $h = 0.75$. The inset shows the change of the isotope composition in the basin ($\delta_y - \delta_p$) as a function of the fraction of re-evaporated precipitation.

Chapter 10

Water in Soils and Plants

10.1. Infiltration, percolation and interflow (*)

As soon as water accumulates on the surface, whether as the result of
direct rain deposition or overland flow onto surface depressions, it infiltrates the top soil layer into the voids in the soil structure. The rate of
infiltration ranges from less than 0.5 mm/hour over heavy (clay) soils to up
to 100 mm/hour on a sandy terrain, on the average. When the precipitation
rate exceeds that of the infiltration, the excess can accumulate in surface
depressions or impoundments which are then emptied through continuing
infiltration and evaporation. Especially during warmer months or when the
incoming waters are irrigation waters during dryer periods, these waters
are often enriched in the heavy isotopes prior to infiltration as a result of
partial evaporation. Based on a simplified model as presented in Box 10.1,
enrichments of up to 5‰ in $\delta(^{18}O)$ can thus be imprinted on the infiltration
flux. In practice, enrichments of a few permil relative to the precipitation
input have indeed been recorded under semi-arid climate conditions (Gat
and Tzur, 1967).

Because of the impaction of the surface, the rate of infiltration is often
less than the amount needed to satisfy the hydraulic transmissivity of the
bulk layer. On the other hand, part of the infiltration flux can occur by
means of cracks or large-pore conduits as a *by-pass flux* to the *matrix
flow*. This is conditional on the occurrence of ponding and thus the by-
pass flows favour the larger rain events. In the extreme case of very intense
rains that exceed the holding capacity of the surface reservoirs, the excess

---

*Compare the entries "Water movement in unsaturated soils" and "Infiltration" by
J. R. Philip in *Encyclopedia of Hydrology and Water Resources* Ed. R. Hershy), 1998,
pp. 699–706 and 418–426, respectively.

Box 10.1. Pre-enrichment of the infiltration flux.

As rainwater accumulates in surface depressions following a rain event, it is exposed to competing fluxes of infiltration (I) and evaporation (E). The latter causes some isotope enrichment in the remnant surface water which is then partially transmitted into the infiltrating waters. The average increase in ^{18}O concentration relative to the initial concentration in the rain (namely $\Delta = (\delta_I - \delta_0)$ as a function of the ratio of I/E was calculated by Gat and Tzur (1967) based on the Craig-Gordon evaporation model. The average enrichment over the whole period until the surface water accumulations are dried up is given as[†]:

$$(\delta_f - \delta_0) = \int (\delta_f - \delta_0) \cdot df \Big/ \int df = \int \{(h\varepsilon^* - \varepsilon)/h\} \cdot (1 - f^u) \cdot df$$
$$\approx (h\varepsilon^* - \varepsilon)/\{1 + (1 - h) \cdot I/E\}$$

where f is the fraction of the water remaining in the surface pools and the integration is performed over the range of f = 1 to f = 0, assuming a humidity value of 75% (h = 0.75) and that the isotopic composition of the vapour is in equilibrium with that of the precipitation (i.e. of δ_0).

The results shown in Fig. A are obtained for typical winter and summer evaporation rates, as indicated. It is noteworthy that the original size (depth) of the water body does not appear as a parameter, because the longer contact time of the larger pools with the atmosphere is being compensated by the larger volume of water that is affected.

[†]Note that in the original paper Eq. (3) was in error and should read as follows:

$$(\delta_f - \delta_0) = (f^u - 1) \cdot (h\varepsilon^* - \varepsilon)/h.$$

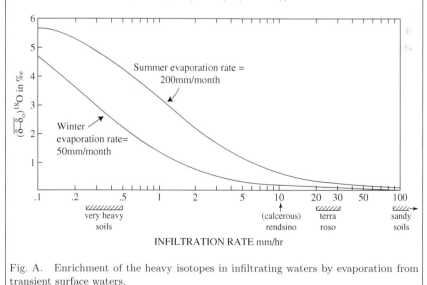

Fig. A. Enrichment of the heavy isotopes in infiltrating waters by evaporation from transient surface waters.

water runs off on the surface and thus contributes only in part to the local infiltration flux.

It is to be noted that the selection between rain events having different isotope values contributes to the change of the isotope composition during the infiltration process.

Water in the soil consists of a part absorbed on the mineral surface and held against gravity in the pores by capillary forces. Additional water percolates through the maze of the soil along a tortuous flow path, predominantly downward but often also as interflow following the morphology of the soil layer. The percolation flux acts essentially as a piston-displacement along the soil capillaries, so that any change in the isotopic composition of the input waters with time (most noticeably the seasonal cycle) is gradually displaced downward and can be traced in the vertical structure of the isotope composition of the water in the soil layer as shown in the example in Fig. 10.1. However, as the water moves downward, it exchanges and mixes with the waters held on the surface of the soil pores. Furthermore, due to the heterogeneity of the pore space and the by-pass fluxes along fissures in the unsaturated soil zone, the variability in the input waters is attenuated with depth.

When the percolation of the excess waters has ceased during the interval between rain events, further changes in the isotopic composition of the water along the soil column are the result of diffusion in the liquid and soil-air layers as well as due to evaporation from within the soil column, as discussed in more detail in Section 10.2.

A special case is presented when the precipitation input is in the form of snow, which blankets the surface until melting sets in. Two situations are to be distinguished relative to the subsequent infiltration of the meltwater:

(i) when the soilwater is in the frozen state (permafrost), most of the winter precipitation is lost in the form of surface flow as melting occurs, because deeper infiltration is prevented by the frozen condition of the sub-surface (Sugimoto *et al.*, 2003);

(ii) when the soil is permeable as melting occurs, a larger fraction of the winter precipitation is often represented in the infiltration flux than if that precipitation fell as rain, since the melting process is delayed and gradual.

As was described in Chapter 7, some isotope change occurs in the snow pack prior to its total meltdown, thus imparting a distinguishable isotope signature to this infiltration mode.

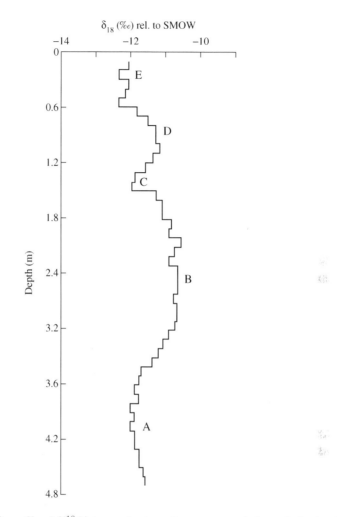

Fig. 10.1. Depth profile of $\delta(^{18}O)$ in a soil where the mean percolation rate is about 260 mm/year. A,C and E correspond to identified winter precipitation whereas D and B are related to input during the summer months.

10.2. Evaporation from within the soil

Evaporation of water from within the soil column differs from that of open waters in that mixing within both the liquid and gas phase is restricted by the texture of the soil matrix. This manifests itself in a large concentration gradient being established within the soilwater column on the one hand,

and that the isotopic compositions of the residual water in the soil establish slow-slope *Evaporation Lines* when plotted on a δ-*plot*.

Let us distinguish between the saturated and unsaturated situations in the soil column. In the saturated soil zone, such as can occur immediately following a precipitation event or in the capillary-rise zone, there is established an isotope concentration gradient between the waters enriched in heavy isotopes at the evaporating surface (termed the *Evaporation Front*) and the water in the deeper soil layer which is close to the undisturbed mean composition of the input water. The shape of such gradients, an example of which is shown in Fig. 10.2, are the result of the diffusive dissipation of the enrichment at the surface and depends on the soil texture and tortuosity. The gradient can be further modified by an advective upward stream of deep soilwater to the surface. In the upper soil layer above the *Evaporation*

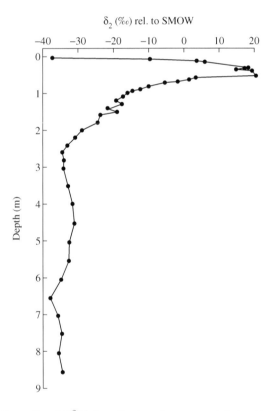

Fig. 10.2.   Depth profile of $\delta(^2\mathrm{H})$ in soilwater of an unvegetated arid-zone sand dune, showing the location of the *Evaporation Front* at a depth of close to half a meter.

Front, the transport is essentially one of vapor transport while below this front, the transport is predominantly in the liquid.

As the evaporation front recedes from the surface, the evaporation rate is reduced due to the lengthening vapour pathway through the dry, unsaturated top soil layers. This process is accompanied by a transition of the slope of the evaporation lines in δ-*space* from that of a free-surface system to a fully developed diffusive boundary layer, where the value of **n** approaches the value of n $=$ 1 (*vid*. Eq. (4.10a)). This has been demonstrated in a series of laboratory evaporation experiments by Allison, Barnes and Hughes

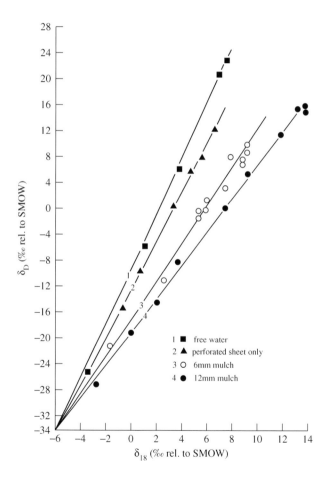

Fig. 10.3.   Evaporation from covered surfaces — based on Allison *et al.* (1983).

(1983), where the evaporating surface was covered by a mulch layer of increasing depth, resulting in the enrichment curves shown in Fig. 10.3.

In the unsaturated part of the soil, water is transported in parallel in the liquid and vapour phase, often along separate pathways due to the heterogeneity of the soil texture. The role of the vapour phase can be one of the homogeneisation of the isotope composition of the more stationary liquid, by means of continuous isotope exchange along the pathway.

A detailed review of the subject was given by Barnes and Allison (1988).

10.2.1. *The soil-mediated recharge flux*

As was described in Section 8.1.3, the isotope composition of the percolation flux from the surface usually is the amount weighted composition of the precipitation events that remain to percolate downward after accounting for the loss of water by direct surface runoff, or evapo-transpiration losses from the top soil layers. To a first approximation, any change in isotope composition between that of the precipitation and the recharge flux is then the result of selection between the precipitation events, based on either the amount, seasonality or phase change effects, and is characterized by a displacement on the δ-*plot* along the *Local Meteoric Water Line*.

Under temperate and tropical climate regimes, any evaporative enrichment of the isotopes in soilwater may be apparent in the water taken up by plants during drier periods, but its impact on the groundwater recharge is negligibly small. However, in dry climate zones, the direct groundwater recharge (in distinction from the surface-water mediated recharge) can be recognized by its isotopic signature along low-slope evaporation lines, as exemplified in Fig. 10.4. An interesting case has been described in shallow groundwaters in the Gobi Desert by Geyh, Gu and Jaekel (1996), Fig. 10.5, where the isotope composition in different wells in the Gurinai oasis appear to prescribe a line parallel to the local MWL with a very negative d-excess value of about $-22‰$. One possible explanation is that these waters represent mixtures of enriched surface waters (related to the precipitation input by an EL of slope ≈ 5) and enriched soilwaters (lying on an EL of slope ≈ 2.5).

10.3. **Water uptake by plants**

To a good approximation, the uptake of water by the roots is not accompanied by any marked change in the isotope composition. The same is

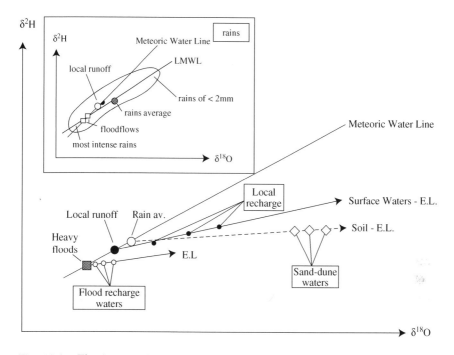

Fig. 10.4. The isotope signatures imposed on arid-zone groundwater by different recharge pathways. The inset on top shows the distribution of the isotopic composition among rains of varying intensities.

essentially true for the further transport as xylem through the stem and branches to the leaves, except for some interchange with the phloem in some plants (Yakir, 1998). Indeed, this conservative property is being used to specify the origin of the water taken up by the plants, either the site of water extraction in the soil or from ground or surface waters, by comparing the stem-water's isotopic composition with that of environmental waters. Among many examples is the identification of the water uptake from the soil profile versus groundwater in an arid environment (Adar *et al.*, 1995) based on the more enriched evaporative signature in the soil water's isotope composition as compared to the local groundwaters. Figure 10.6 shows the composition of water absorbed by different plants growing in the Negev Highlands, obviously a function of the soil water composition and depth of their root system. Another example, first described by Dawson and Ehleringer (1991), traced the origin of water taken up by trees that grow next to a river to the locally infiltrating rainwater rather than the river

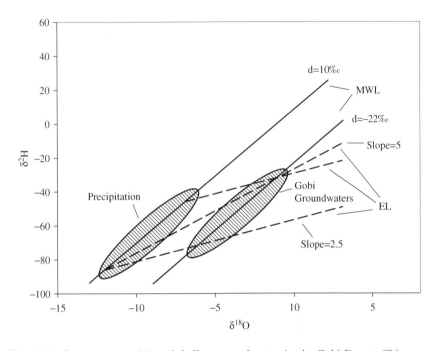

Fig. 10.5.   Isotope composition of shallow groundwaters in the Gobi Desert, China, as reported by Geyh, Gu and Jaeckel (1996) and their assumed formation from the local precipitation following evaporative processes.

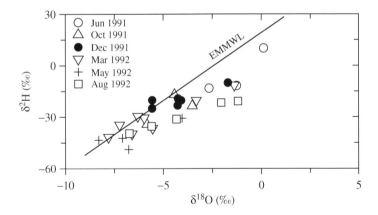

Fig. 10.6.   Water isotope composition extracted from the stems and branches of plants growing in the Negev highlands. EMMWL signifies the Local Meteoric Water Line.

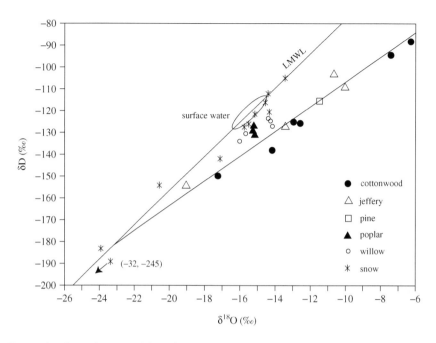

Fig. 10.7. Isotopic composition of water extracted from tree-stems and twigs growing in the vicinity of a creek in the Sierra Nevada, USA. The Local Meteoric Water Line, the isotope composition of the surface waters in the creek and local snowfall are shown for comparison.

water. Figure 10.7 from a study of Kaplan *et al.* (1995) in the Sierra Nevada shows a similar pattern but also emphasises the difference between the different tree species; only the willow trees and poplars evidently utilise in part the river waters, which at this particular location are easily distinguishable from the local soilwater which is mainly replenished by snow-melt from the winter precipitation. Additional examples were reviewed by Dawson *et al.* (1998).

Yakir and Yechieli (1995) investigated plants growing in a hyper-arid environment in a dry-river (wadi) bed, where it could be shown that the plants preferentially took up water from freshwater pockets remaining after inundation by a flood. A special case which could be identified by this method is that of the uptake of fog-drip by the plants, on the basis of the very different isotope composition of these waters when compared to the "evaporated" soil waters, e.g. in northern Kenya by Ingraham and Mathews (1988) and in the Redwoods of California by Dawson (1998).

10.4. Water in the plant tissues

Gonfiantini *et al.* (1965) first reported the enrichment of heavy oxygen isotopes in the leafwater of plants, presumably as a result of the evaporation from the leaves. Wershaw *et al.* (1970) reported similar results for the isotopes of hydrogen. Modeling the evaporation of water from leaves as that of a *terminal evaporation system* and on applying the Craig-Gordon evaporation model (*vid.* Section 4.3) one would expect the buildup of isotopes to be given by Eq. (10.1):

$$d(V \cdot \delta_{LW})/dt = F_{in} \cdot \delta_{SW} - E \cdot \delta_E \qquad \delta_E \approx (\delta_{LW} - h \cdot \delta_a - \varepsilon)/(1 - h);$$
$$\varepsilon = \varepsilon^* + \Delta\varepsilon; \quad \Delta\varepsilon = (1 - h) \cdot C_k \tag{10.1}$$

Under both hydrologic and isotopic steady-state conditions, this yields the following value for the isotope composition of the leaf-water ($\delta_{LW,ss}$):

$$\delta_{LW,ss} = \varepsilon + h \cdot \delta_a + (1 - h) \cdot \delta_{SW} \tag{10.2}$$

This relationship applies to both the hydrogen and oxygen isotopes, with the relevant fractionation factors for the isotopic species concerned.

On the average, a steady-state relationship appears reasonable since the transpiration flux exceeds the hold-up volume in the leaf manifolds. Under such conditions, the isotopic composition of the transpiration flux (TF) should then match that of the source water (SW) taken up by the plant, i.e. $\delta_{in(SW)} = \delta_{out(TF)}$.

When both the hydrogen and oxygen isotope enrichments were compared in the leaf water, among others by Bricout *et al.* (1972), Allison *et al.* (1985), Yakir *et al.* (1990) and Nissenbaum *et al.* (1974), the isotope compositions of the leafwater were found to be situated off the Meteoric Water Line (MWL), generally aligned in δ-space along lines with a slope of less than 8, as befits residues of evaporated waters. In most cases, these "Leafwater Lines" originate from the relevant composition of the source waters taken up by the plants. Their slope, i.e. $S = \Delta\delta(^2H)/\Delta\delta(^{18}O)$, being smaller than the corresponding ones of surface waters evaporating under similar climatic conditions, was explained as due to the fact that evaporation of the leaf-water occurs from within the stomata and through a stagnant air pocket, namely through an air layer wherein the diffusive fractionation between the isotopologues is fully developed, as is the case also for the evaporation of soilwater. In support of this hypothesis, observe

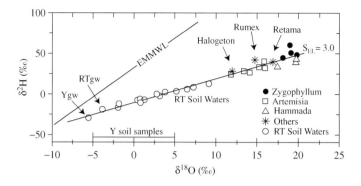

Fig. 10.8. Isotope composition of leafwater of plants growing in the Negev Desert high-lands. EMMWL signifies the eastern Mediterranean Meteoric Water Line with d = 20‰; Y_{gw} and RT_{gw} are local groundwaters.

Fig. 10.8 for data of water from plants growing in the Negev Desert high-lands; these are seen to be situated along the extension of the *Evaporation Line* of the soilwaters in that region.

A closer look at the leafwater's isotope composition found in these and similar studies reveal additional features that need to be addressed, namely:

- the cases where the isotope composition of water extracted from the leaf falls short of the expected steady state composition as defined by Eq. (9.2) for a terminal water body undergoing evaporation.
- large changes in the isotopic composition during the diurnal cycle and under different water stress conditions in a single plant species as well as differences in the build-up of the evaporative signature in various plant species under identical environmental conditions, as exemplified in Fig. 10.9.
- the case of *"Leafwater Lines"* that do not extrapolate back to the origin of the source waters, for example as shown by Allison *et al.* (1985) and discussed below.
- differences in the isotopic composition of water in different parts of the plant. Examples of this are between the bottom and top of long leaves, as reported by Wang and Yakir (1995), or between leaflets and petioles in the desert shrub *Zygophylum Dumosum*, as shown in Fig. 10.10.

The first of these observations has been explained by the fact that leafwater is not a single (mixed) pool of water, but is compartmentalised, with a limited exchange between the water pool directly exposed to water loss by transpiration with the symplastic water in the cells wherein the

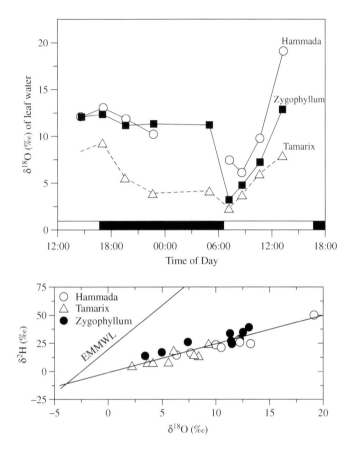

Fig. 10.9.   The diurnal cycle of the isotope composition of different plants growing under water stress in a desert area. The lower plot shows these data on a δ-plot.

photosynthetic reaction takes place. Water in veins conducting water to the leaves, containing unchanged feed water, constitutes a third compartment. In a study on ivy and sunflowers, Yakir *et al.* (1990) estimated that the leafwater was constituted of roughly 69%, 29% and 2% of these three water fractions, respectively. Obviously, these numbers differ for different plant species as well as during the diurnal cycle, as more and more of the water is evaporated while the stomata are open and when the water deficit is later replenished by unfractionated source water. The timing and degree of opening of stomata is a key factor in this.

Another factor that needs to be taken into account when interpreting the data is the differences in the environmental conditions, in particular the relationship between the isotopic compositions of the source waters

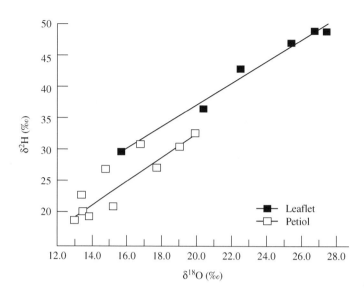

Fig. 10.10. Isotope composition of water in different parts of the desert plant *Zygophyllum Dumosum*.

to that of the ambient moisture in the atmosphere. In this respect, the mixing characteristics of the near-surface air above or within the canopy is also of concern, as it controls the admixture of the transpiration flux within the atmospheric moisture and thus, ultimately, the humidity gradient within the boundary layer. It is evident from Eq. (10.2) that the expected enrichment as well as the slope of the *Evaporation Line* depends on the humidity value, as shown in Eq. (10.3).

$$S = (\delta_{\text{LW,ss}} - \delta_{\text{SW}})_{2\text{H}} / (\delta_{\text{LW,ss}} - \delta_{\text{SW}})_{18\text{O}}$$
$$= [h \cdot (\delta a - \delta_{\text{SW}}) + \varepsilon]_{2\text{H}} / [h \cdot (\delta a - \delta_{\text{SW}}) + \varepsilon]_{18\text{O}} \qquad (10.3)$$

As discussed in more detail in Chapter 9 and based on Gat (1995), the slopes of the relevant evaporation lines are different for the different humidity values, with the exception of the case when the isotope composition of both the inflow water and the atmospheric humidity are in isotopic equilibrium with respect to the liquid to vapour transition; in this case, all these *Evaporation Lines* merge into one whose slope is not any more a function of humidity and is given by the equation:

$$S = (\varepsilon^* + C_k)_{2\text{H}} / (\varepsilon^* + C_k)_{18\text{O}}. \qquad (10.4)$$

The steady state values given by Eq. (8.2) for various humidities form a locus line whose equation is:

$$\delta_{\mathrm{LW,ss}} = B \cdot h + A \qquad (10.5)$$

where

$$A = (\delta_{\mathrm{SW}} + \varepsilon^* + C_k) \quad \text{and} \quad B = (\delta a - \delta_{\mathrm{SW}} - C_k).$$

Unlike the *Evaporation Lines*, this *Steady-state Locus Line* does not incorporate the isotopic composition of the input waters, except for the special case of equilibrium between the isotope composition of the input waters and the atmospheric vapour. Since the humidity changes during the day and as long as stomata are open, the isotope composition of the leafwater can then be expected to move along such a locus line. Consistent with this view are the two cases shown in Fig. 10.11: the line showing

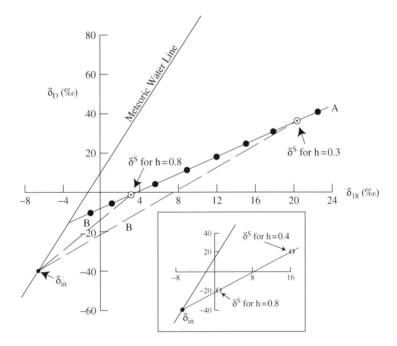

Fig. 10.11. Schematic representation of isotope enrichment in plant material expected under steady-state conditions according to Eq. (10.5). Line A represents the locus line, to be compared to the individual Evaporation Lines, B, shown for 2 cases, namely for h = 0.3 and h = 0.8; δ_{in} situated on the MWL represents the feed water. The inset shows the case where the ambient air is in isotopic equilibrium with the inflow water.

the data from Europe are from an area where usually the ambient air is in isotopic equilibrium with the precipitation (as discussed in Chapter 6) whereas in the arid climate of southern Australia, especially in a coastal setting, such an equilibrium is not the rule.

So far, the evaporation was viewed as occurring from a single terminal system fed by unfractionated input water. However, the examples cited above by Wang and Yakir and in Fig. 10.11 indicate that a series of successive coupled evaporation sites may be a more appropriate description of the system. The enrichment can then be described by the string-of-lakes terminology (Gat and Bowser, 1991) or a more elaborate two-way flow scheme as recently discussed by Cuntz *et al.* (2007).

While on the whole, the assumption that $\delta_{TF} = \delta_{SW}$, can stand, where δ_{TF} signifies the isotope composition of the transpiration flux and δ_{SW} that of the source water, this does not hold exactly throughout the diurnal cycle, especially in the early morning hours when the leafwater has not yet reached a steady state with respect to the evaporation process.

Comprehensive reviews of the isotopic composition of plant material were given by Yakir (1998), Farquhar *et al.* (1998) and Dawson *et al.* (1998) in the publication *"Stable Isotopes: integration of biological, ecological and geochemical processes."*

Chapter 11

Saline Waters*

In classifying saline waters, both the total salinity (TDS) and the chemical composition has to be recognised. In view of the ubiquitous prevalence of seawater and its derivatives, it is common to distinguish between brackish waters and brines, where the former term applies to salt concentrations in between freshwaters and marine waters, and the term brine refers to concentrations beyond that of the ocean water. With respect to the chemical composition, both the anionic and cationic dominances are taken into account, as pictorially portrayed on the triangular diagrams, as exemplified in Fig. 11.1 for water sources of the Sinai Desert. A full description of the pictorial representations of the chemical composition was given in "The Properties of Groundwater", by Matthess and Harvey, Wiley-Interscience.

A variety of measures of the salinity are prevalent in the literature. Commonly, one defines salinity (S) as the weight in grams of inorganic ions dissolved in 1 kg of the solution. The units used are thus "parts per thousand (ppt" or "‰). It is to be noted that this measure refers to 1 kg of the solution, not to 1 kg of the water substance; as will be discussed, this issue will arise when considering the change in isotopic composition of mixtures of waters of different salinities.

In seawater and derived saline waters, in which the chemical composition is dominated by NaCl, the measure of the chloride content (Cl as g/kg) is a more practical alternative, and as discussed by Defant (1961), it can be related to the salinity by the empirical relationship: S = 0.030 + 1.8050·Cl.

---

*See also the review by J. Horita: Chapter 17 in *Isotopes in the Water Cycle*, (P.K. Aggarwal, J.R. Gat and K.F.O. Froehlich, eds), Springer, 2005.

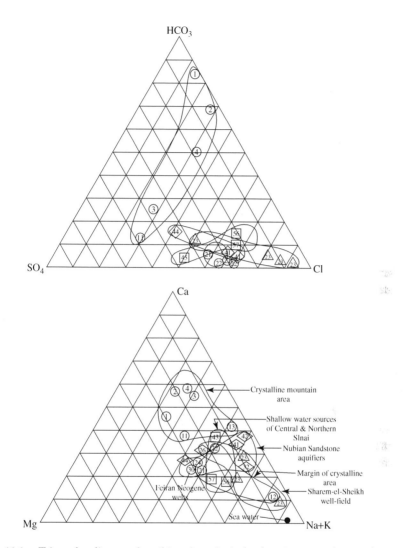

Fig. 11.1.   Triangular diagram describing the anionic (top) and cationic (bottom) composition of water sources of the Sinai desert (Issar and Gat, 1974).

The measure of salinity by volume, in terms of gram of salts per liter of the solution (g/L), is not to be recommended unless the density of the brine is well defined both with regards to temperature and salt content.

One can clearly distinguish between saline waters originating from fresh waters of the meteoric water cycle and those derived from seawater based on their isotopic composition, as is discussed in Section 11.1.

The analytical aspects of isotopes in saline waters were outlined in Box 2.2.

11.1. Sources of salinity in the hydrologic cycle

The primary source, the ocean, with a range of salinity of S = 32–37‰ and a mean value of 35‰, is accompanied by isotopic values ranging by about ±2‰ in $\delta(^{18}O)$ around SMOW (Magaritz and Gat, 1981). In surface waters where evaporation exceeds both precipitation and runoff from adjoining continental areas, the heavy isotope species are slightly enriched; Craig (1965) estimated a mean value of $\delta(^{18}O) = +0.5$‰ for oceanic surface waters. The highest values are on record for semi-enclosed basins in arid zones such as the Persian Gulf and the Red Sea, where Craig (1966) reported values of $\delta(^{18}O) = +2$‰ for a salinity of S = 42‰. The commensurate enrichment of the hydrogen isotopes occurs along an Evaporation Line of slope $\Delta\delta(^2H)/\Delta\delta(^{18}O)\approx6$. In contrast, low δ-values are characteristic of ocean masses in estuaries or close to the outflow points of large river systems. As was shown in Fig. 5.2 by Friedman *et al.* (1964), the isotopic compositions then form linear mixing lines between the ocean composition and the relevant isotope composition of the freshwater source on a plot of $\delta(^{18}O)$ versus the salinity.

The marine waters appear as primary sources of the salinity in coastal lagoons and inland brine accumulations, where they were often apparently introduced during past interglacial epochs when the sea level was somewhat higher so that coastal and inland low-lying areas were inundated. It should be mentioned, as was briefly discussed in Chapter 5, that the isotopic composition of the ocean waters at that time were somewhat depleted relative to SMOW.

Aeolian transported salinity, although also primarily of marine origin, especially in near-coastal settings, is accompanied by the isotopic composition of the meteoric waters as it is deposited by the precipitation. It can then be modified by interaction with the soil and aquifer material and by dissolution of evaporitic sediments such as gypsum. The accompanying change in the isotope composition of the water is usually minor (except under high temperature situations), and affects only the oxygen isotopes in the case of interaction with carbonates or silicates. Dissolution of hydrated minerals such as gypsum, on the other hand, affects both hydrogen and oxygen isotopes due to the isotope composition of the crystal water in these minerals, as discussed in Box 11.1.

Box 11.1. The isotope composition of hydration water.

Taube (1954) who investigated the structure of water molecules that are held in the hydration sphere of ions, discovered that the isotope composition of the water in the hydration sphere of certain ionic constituents differed from that of pure water. Taube used the CO_2 equilibrating technique, comparing the isotope composition of oxygen in carbon-dioxide equilibrated with pure water to the values obtained in aqueous solution of different salts. It was found that the heavy oxygen content of the CO_2 decreased upon addition of $MgCl_2$, $AlCl_3$, HCl or LiCl, remained essentially unchanged for Na salts, but increased when CsCl was added. These changes are roughly proportional to the molality of the solution and interpreted as pointing to a different structure of water in the hydration sphere. Terms such as "structure-making" and "structure-breaking" effects of the ions were invoked to explain these findings, as reviewed by O'Neil and Truesdell (1991). Sofer and Gat (1972) confirmed and extended these findings to other salts, with a negative effect shown also for K ions (Fig. A). It was further shown that in the range of concentrations of natural brines these effects are additive, and the following equations were given for the correction to be applied to the measured δ_m value relative to δ_0: {i.e. $\Delta\delta^* \equiv (\delta_m - \delta_0) / (\delta_m + 1000)$}. (concentrations to be expressed in molalities)

for ^{18}O: $\Delta\delta^* = 1.11 M_{Mg} + 0.47/M_{Ca} - 0.16\ M_K$, ($\sim 0$ for NaCl and anions)
for ^2H: $\Delta\delta^* = -6.1\,M_{Mg} + 5.1\,M_{Ca} + 2.2\,M_{Na}$ (for chlorides, there is an anion effect)

One notes the opposite salt effect for the divalent ions for the two isotopes. The relatively small effect of NaCl justifies ignoring the salinity correction in the determination of the isotopic composition of sea water.

The hydration water in minerals such as gypsum {$CaSO_4.2H_2O$} was found to be correspondingly enriched in ^{18}O and depleted in Deuterium relative to the mother liquor under equilibrium conditions (Fontes, 1965; Fontes & Gonfiantini, 1967; *vid.* Table 3.1). Natural gypsum deposits were found, however, to cover a wide range of isotopic compositions, extending on both sides of the Meteoric Water Line, as shown in Fig. B. This was explained by Sofer (1978) by two scenarios for forming these deposits: from an evaporating brine with its enriched

(*Continued*)

Box 11.1. (*Continued*)

δ-values (Line a with slope of S $= \Delta\delta(^2\text{H})/\Delta\delta(^{18}\text{O})=$ 2.6 to $+5$), and from anhydrite {CaSO_4}deposits invaded by percolating water, whose isotopic composition then changes along Line b with a slope of S $= -5$.

Obviously, when such gypsum deposits are then dissolved in freshwaters, the incorporation of the crystal water into the water pool imparts an identifiable signature as shown in Figs. 11.3 and 11.4.

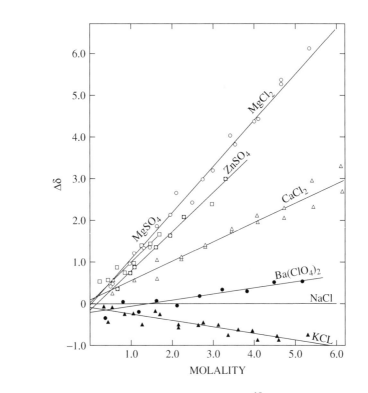

Fig. A. Magnitude of the isotopic correction term $\Delta\delta^{18}\text{O}$ as a function of molality in various aqueous salt solutions.

(*Continued*)

Box 11.1. (*Continued*)

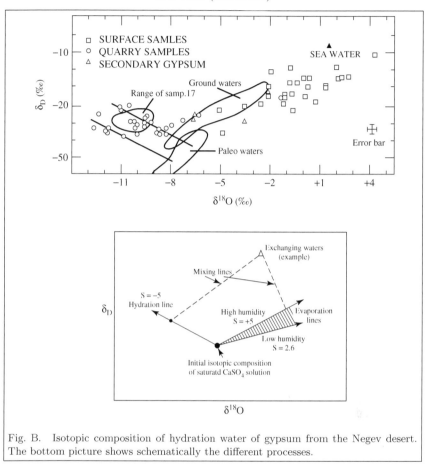

Fig. B.   Isotopic composition of hydration water of gypsum from the Negev desert.
The bottom picture shows schematically the different processes.

11.2. Salinisation mechanisms and their isotopic signature

As outlined in Table 11.1, a variety of processes result in the buildup of
salinity in the water bodies as they move through the surface or subsurface
systems. One can distinguish between processes adding salts to the waters
and those that concentrate and/or modify the existing saline burden.

The change in isotopic composition as a result of these processes is
shown schematically in Fig. 11.2 as a function of the salinity.

The leaching or pickup of salinity from the surface, soils or geologic
formation is not accompanied in most cases by an isotopic effect, except

Table 11.1.   Processes resulting in salinity increase.

1. Salt addition:
    A_1— atmospheric deposition (dry and rainout)
    A_2— Flushing of accumulated salinity (surface or soils)
    B — Dissolution of salts
    C — Mixing with brines

2. Concentration of existing salt burden:
    D — Evapo-transpiration
        D_1 — Evaporation from open surfaces
        D_2 — Evaporation from soils
        D_3 — Transpiration
    E — Partial freezing
    F — Ultra-filtration

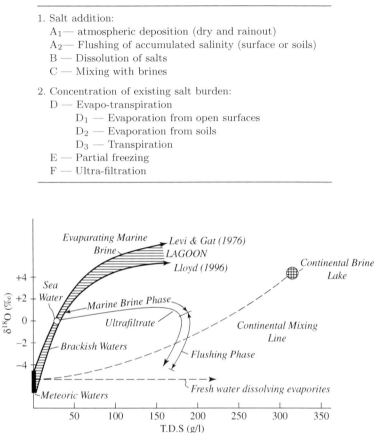

Fig. 11.2.   The predicted evolution of $\delta^{18}O$ vs. salinity for various salinisation processes.

in the case of hydrated minerals whose isotopic composition is discussed in
Box 11.1, or in the case of remnants of an antecedent evaporitic process,
as in the case of leaching of soil water by percolating water, when the
percolating waters mix with the "enriched" residual soil water. When the
salinity results from the admixture of a brine with fresher water (process C,
Table 11.1), the resultant isotope composition is a simple linear mixture of
their δ values, provided the δ values are given on the concentration scale
(*vid.* Box 2.2). As shown on Fig. 11.2, these mixing lines are not linear,
however, at high salinities on a plot of δ versus salinity, except when salinity
is expressed in *molal* rather than the usual concentration units.

11.2.1. *Evaporation of saline waters*

Unlike evaporation from a body of fresh water, where the heavy isotope content builds up until it reaches a steady state value, equivalent experiments on the evaporation of seawater and other saline waters showed that the isotope enrichment was often reversed when higher salinities accumulated from the evaporation.

To a first approximation, the stable isotope balance and fractionation is not much affected by the salinity in brackish and moderately saline solutions. When concentrations exceed those of seawater appreciably, however, a number of "salinity effects" on the isotope fractionation assume importance. Considering the enrichment process as described in Table 9.1, it is foremost the effect on the humidity term, 'h' that counts. Obviously, 'h' reaches higher values over brines when compared to non-saline waters for any given atmospheric vapour pressure, since the humidity term in these equations is normalised relative to the saturated vapour pressure over the solution, and the latter is reduced with increasing salt concentration, as first reported by Gonfiantini (1965). As shown in Fig. 11.3, the enrichment is reduced relative to a fresh water body evaporating under similar conditions until evaporation ceases when the reduced vapour pressure over the saline solution equals that of the ambient humidity. Following that, exchange between the liquid and the ambient humidity can then continue

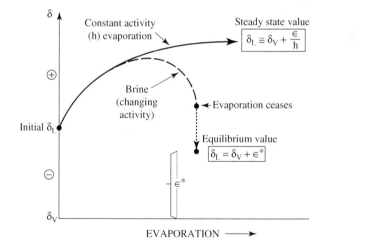

Fig. 11.3. The change of the isotope compostion in an evaporating water body as salinity builds up.

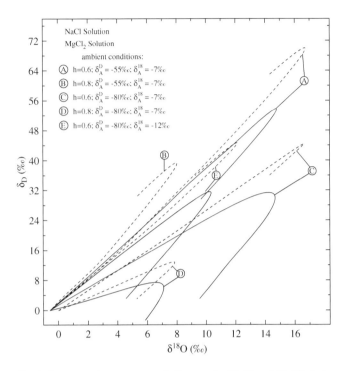

Fig. 11.4.   Typical evaporation curves on δ-diagram for NaCl and MgCl$_2$ brines of initial concentration of 0.5 molal. Based on Sofer and Gat, 1975.

until isotopic equilibrium is established between them. Furthermore, the specific salt effect on the equilibrium isotope fractionation factor has to be considered; this effect results from the preferential binding of one or the other isotopic species in the hydration sphere of the ions concerned. As shown in Fig. 11.4, on a δ-plot, the evolution of the isotopic composition with increasing salinity in such evaporating systems differs then for different ionic makeup of the solution. It is further to be noted that when the evaporating system reaches the solubility limit of its salty constituent, a new steady state enrichment is established. In this respect, one needs to distinguish between a salt such as halite (NaCl) and a mineral such as gypsum (CaSO$_4$·2H$_2$O), which incorporates water of crystallisation.

11.2.2.   Freezing of saline solutions

When a saline solution freezes slowly under equilibrium conditions, the salinity is not incorporated into the newly formed solid phase and accumulates in the remnant liquid (indeed this process has been suggested as a device for desalinisation of seawater, by the Zarchin-Process). The isotopic composition then changes in accordance with the equilibrium fractionation factor for the liquid/solid phase transition, as given in Table 3.1. The slope of the line for such a process on a δ-diagram, closely resembles a typical slope of the *Evaporation Lines*. However, whereas the evaporation loss leaves the liquid enriched in the heavy isotope content, the opposite sense is imposed on the freezing process, as was already discussed in Section 9.2.3. However as shown in Box 9.1, the actual freezing process often does not take place under complete equilibrium conditions but includes the trapping of part of the solution in the growing ice matrix without any isotope fractionation or exclusion of the salt content.

11.2.3. *Ultrafiltration*

Another process whereby the salt content increases in the water body occurs when a solution diffuses through a semi-permeable clay layer under a pressure head. This so-called ultra-filtration process (Coplen and Hanshaw, 1973) results in the accumulation of most salts, especially Calcium and Magnesium salts, and is accompanied by a slight enrichment of both oxygen and hydrogen isotopes in the remaining saline waters, albeit to a smaller extent than most other processes and along a slope favoring the enrichment of oxygen isotopes over those of Hydrogen, as was shown in Fig 11.2, which summarises the isotope enrichment of the different processes enumerated above.

Chapter 12

Sub–Surface Waters

Water is encountered beneath the surface within the lithosphere in a variety of hydrologic contexts. The upper soil, or rock zone, where the voids in the structure are only partially saturated with water and the *vadose* zone, have been discussed in Sections 9.1 and 9.2. In it, the water movement is essentially in the vertical, consisting of the downward drainage of excess water beyond the holding capacity of the soil and an upward flux of evaporating moisture. As the water accumulates on an impermeable formation (*aquiclude*), the voids become filled (i.e. *saturated*) with water forming the so-called "groundwater" bodies. One then can distinguish between small local accumulations of water within the vadose zone (*perched groundwater*) whose overflow rejoins the downward percolation flux and the large regional groundwater bodies which drain through the *aquifers* in response to the pressure gradients, finally to emerge downstream as springs or as coastal and bank seepage into surface water bodies. On rare occasions, especially in karst terrain, one encounters underground pools of water or subsurface lakes.

In deeper formations, one can encounter stagnant groundwater which is disconnected from actual recharge and drainage. Such a situation occurs when the recharge flux has been curtailed or stopped by either climatic change or environmental changes of the replenishment regime at the recharge sites. Prominent examples are the *palaeo-waters* under the Sahara, whose waters derive from recharge during the wetter period of the Pleistocene.

Other fossil groundwaters are the *formation waters*, which are mostly remnants of oceanic interstitial water in uplifted marine formations; these obviously are quite saline.

In general, in temperate and humid climates, the stable isotope composition of the groundwater rather closely matches that of the precipitation on the recharge areas, except for cases where recharge from lakes or other surface bodies which have undergone evaporation contributes significantly. The distortion of the isotope composition is small and both precipitation and the recharge flux obey the Meteoric Water relationship (Craig, 1961). There is, however, the possibility of a seasonal shift along the MWL. The high recharge rates result in continuous flushing of remnant waters not related to the meteoric water cycle.

In semi-arid regions, one notes the growing importance of evaporation losses before and during recharge, resulting in an enrichment of the heavy isotopes in the groundwaters relative to the precipitation and a shift away from the *Meteoric Water Lines*. However, the local precipitation still dominates the composition of the groundwater.

In the dryland regions proper, where rain is scarce and irregular, direct recharge to groundwater sources is minor and occurs mainly by means of floodflows through channel bottoms. Water unrelated directly to the local precipitation then assumes importance. Such waters originate in faraway mountainous fringe areas or might be paleowaters, i.e. waters related to recharge in past climate periods. Non-meteoric fossil brines may then also play a role.

In cold regions, where the recharge areas are blanketed at times by snow and ice cover or affected by permafrost, the relationship between the isotopic composition of the precipitation and the groundwater is dominated by the snow-melt processes.

12.1. Circulating (meteoric) groundwater

The groundwater under the vadose zone, which is in open communication with the atmosphere, is named a *phreatic* or unconfined system, and the interface between the saturated and unsaturated zones is then the water table where the pressure is the atmospheric one. Depending on the pore structure and mineralogical nature of the aquifer, a transition zone can develop above the water table, the so-called *capillary fringe*, where the water is held by capillary forces. The isotopic composition at the *water table* is a mixture of the local recharge flux (*vid.* Section 8.1.3) with the upstream groundwater. Further downstream the isotope composition is continuously modified by additional influx as long as the *phreatic* conditions prevail.

Confined conditions develop during the flow in an aquifer when an impermeable confining layer (an *aquiclude*) intervenes between the groundwater and the upper soil layer. The isotope composition of the water is then modified only by mixing with other flowing subunits, dependent on the hydraulic structure of the aquifer.

Unlike the chemical nature of the groundwater which can be continuously modified by geochemical interactions with the aquifer material, the isotopic composition of groundwater is practically invariant with respect to interaction with the aquifer material during the passage through the aquifer as long as the temperature is less than about $60°C$. The composition can be modified, however, by admixture of other waters, either additional recharge (during the phreatic phase of the groundwater flow) or leakage from adjacent and deeper aquifer units. Two extreme cases of the incorporation of the recharge into the flowing groundwater can be described. The recharge is either mixed locally throughout the aquifer depth or, on the other extreme, continues along a parallel flow path downstream, resulting in a stratified depth structure which is then mixed only at the discharge point. Obviously, the isotope composition in the mixed groundwater is the amount-weighted average of all the inflows.

The main use made of the isotopic signature of groundwater has been the identification of the locality of the recharge (based for example, on the altitude effect on the isotope composition of the precipitation), the seasonality of the recharge (based on the seasonal cycle of the isotope composition of the recharge flux, *vid* Section 7.1.3) or the determination of the amount of recharge by surface waters which are identified by their evaporative signature.

12.2. Paleowaters

Paleowaters (also called Palaeowaters) were defined by Fontes (1981) as waters originating from water cycles under environmental conditions which are different from the present ones. Obviously, one expects to encounter such waters in systems with a very long residence or transit time, especially when these became stagnant due to the curtailment of recharge at present. The classic examples are North African aquifers first described by Muennich and Vogel (1962). Other cases described in the literature include the deep groundwaters under the Sinai and Negev deserts (Issar *et al.*, 1972; Gat and Issar, 1974) and large continental confined aquifers such as those in the Great Artesion Basin of Australia and the Stampiet aquifer in South Africa.

The long residence times of these waters was confirmed by dating, i.e. the absence of measurable Tritium and low ^{14}C levels. The waters under the Sahara, as an example, typically showed ^{14}C levels of less than 10 PMC (Sonntag *et al.*, 1978), clearly relegating the recharge to the Pleistocene or even before. The stable isotope composition of these paleowaters differs from the sparse modern precipitation in these regions, usually showing both more depleted $\delta(^{18}O)$ and lower *d-excess* values. Figure 12.1 (Gat, 1983)

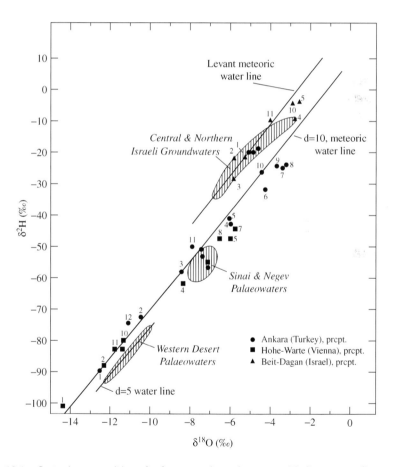

Fig. 12.1. Isotopic composition of paleowaters from the eastern Mediterranean Sea area, compared to modern precipitation in the area (monthly averages based on the GNIP data; numbers signify the respective month).

summarises data from the eastern Mediterranean region on the background of the composition of modern precipitation values, as provided by the GNIP network. It can be seen that the paleowaters are more depleted in both isotopes compared to the local modern precipitation and also aligned along a MWL with a smaller *d-excess* value. While this pattern is repeated to varying degrees in most other cases where paleowaters are compared to modern precipitation in the respective regions, and then interpreted as due to the lower temperatures prevailing at that time, it is to be noted however, that there are exceptions to this pattern in the temperate zones with very small changes in the isotopic composition, for example as described by Vogel (1967) in the Netherlands and by Fontes and Garnier (1979) in the case of the Calcaires carbonniferes aquifer in the north of France.

The isotopic content of the paleowaters is being investigated with two objectives in mind:

- from a water resources management point of view, namely to establish whether there is an additional, more recent recharge. Whereas the exploitation of a non-rechargeable closed-system aquifer is essentially a "mining" operation, the identification of recent recharge would make such waters more amenable to exploitation and enable one to define a "safe exploitation yield".
- From a paleo-climate study point of view, with the aim of using the isotope composition of these waters as *climate proxies*.

The interpretation of the isotope signature of these waters in terms of the paleo-climatic record of the water cycle requires clarification of a few issues:

- to what extent the geochemical and isotopic signature of the recharge waters is conserved during the passage through and the sojourn in the aquifer.
- the extent to which the present hydro-meteorological pattern can serve as guidelines to the interpretation of the paleo-record.
- the effect of the differences in the eco-hydrological character of the transition from precipitation to recharge on the isotope composition, between the present and the past situation.

The chemical character of these "old waters" is modified to some extent by interaction with the aquifer material as shown by the higher salinity of such waters. While it is difficult to asses the accompanying modification of the isotopic signature, the latter apparently is not very large as is evident from the good linear correlation of many of these waters along MWL's. Exceptions to this are cases of very acid conditions or elevated temperatures, as well as situations of admixture of extraneous waters such as saline formation waters.

Concerning the second issue, the most simplistic interpretation of any change in the isotope signature between the present and the paleowaters is based on the temperature dependence of the isotopic composition of the precipitation in the water cycle. This may be misleading, however, considering the possible changes in the meteorological pattern, such as different source regions of the precipitation, changes in the seasonality of the recharge process and the changing role of snow in the process.

In the interpretation of the paleowater data in northern Africa, (compare Fig. 12.1) Sonntag *et al.* (1978) pointed to the close correspondence of the Western (Egyptian) Desert data with the present-day precipitation in eastern Europe (cf. winter precipitation in Ankara), albeit with a reduced *d-excess* value. They suggested a southward displacement during the Pleistocene of the European precipitation regime to the north African continent, acting on an Atlantic moisture source. The relatively less depleted values of the Sinai and Negev paleowaters, which would have been then expected to show even more depleted isotopic values, were interpreted by Horwitz and Gat (1984) as being the result of recharge of summer rains rather than the prevalent winter regime. Obviously, assuming different source regions for the moisture at that time, for example, "monsoonal" rains originating in the Indian Ocean, these data could also be explained as shown in Fig. 12.2, adapted from Gat and Carmi (1987).

When interpreting the isotopic composition of the paleowaters in terms of the assumed isotope composition of the precipitation at that time, it must also be taken into account that the shift in the isotope composition between the precipitation to recharge at any particular site (the *Isotope Transfer Function*) may have altered as a result of the changed environmental conditions. Since the paleowaters mostly derive from more pluvial periods than the present situation, one does not expect a notable evaporative signature in the recharge flux at that time so that the low *d-excess* value of the paleowaters can usually be safely related to the source region of the moisture. However, a seasonal shift in both the precipitation and recharge fluxes compared to the present situation should be considered.

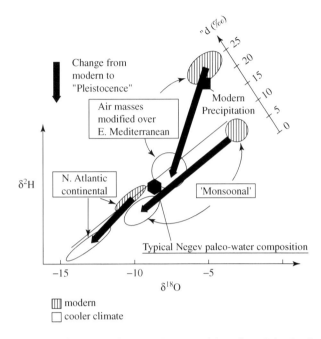

Fig. 12.2.   Postulated changes in the isotopic composition of precipitation in the eastern Mediterranean region originating from either the Atlantic, the Indian ocean or the Mediterranean Sea as possible sources for the paleowaters. Shaded areas represent present-day compositions and the open circles the postulated values for the Pleistocene.

12.3. Geothermal systems

Hot water springs and other geo-thermal occurrences have been used since antiquity. More recently, they have been widely explored and exploited as alternative energy sources all over the globe. Initially, it was believed that the source of water, just like that of the heat, is *magmatic* in origin. However, right from the beginning of isotopic measurements on geo-thermal systems by Craig (1963), a meteoric origin of many of the geothermal springs was inferred. As shown in Fig. 12.3, the isotope composition of the thermal waters from a variety of geographic locations each records more or less the $\delta(^2H)$ of the local precipitation whereas the $\delta(^{18}O)$ values are enriched (shifted), so that the isotopic composition of these waters extend in a series of parallel lines to the right of the MWL on the δ-diagram. This pattern is explained by the high-temperature isotopic exchange of the oxygen isotopes between the meteoric water and the carbonate or silicate formations, whereas the hydrogen isotopes are practically unchanged, with

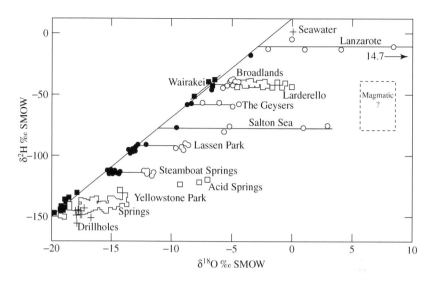

Fig. 12.3. The isotopic composition of thermal waters from a variety of geographic locations on a δ-diagram, showing the Oxygen shift.

the possible exception by interaction with clay minerals or hydrated salt deposits.

Some waters from *fumaroles* and vapour-venting hydro-thermal systems describe a different pattern of alignment along low-slope lines to the right of the relevant MWL (Gerardo-Abayo *et al.*, 2000), an example for which is shown in Fig. 12.4. Initially explained in terms of mixtures of meteoric waters with a hypothetical *magmatic* water (Giggenbach, 1992), Craig, Boato and White (1956) showed that the lines for different hydrothermal sources did not converge on a single hypothetical magmatic source and that it was more reasonable to invoke the high-temperature fractionation during the liquid to vapour transition as the ascending hot waters boil and are vented (Truesdell, *et al.*, 1977; reviewed by D'Amore *et al.*, 2000). Some authors have, however, postulated different deep water sources, resulting from the high-temperature interaction of the magmatic exhalation with the local geological formations. It is to be further noted that the ratio of the fractionation factors for the liquid to vapour transition for Deuterium and Oxygen-18 isotopes, respectively, is also a function of temperature and thus differs somewhat in different systems. It is evident that there is so far no consensus on the evolution of some of the hydrothermal systems.

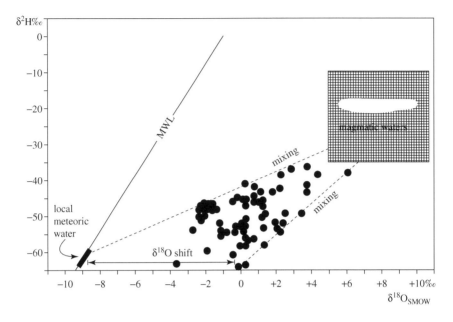

Fig. 12.4. δ-diagram of fumarolic condensate in a hydrothermal system in Japan, explained as the mixing between meteoric waters or Oxygen-shifted meteoric waters and hot deep-seated magmatic waters.

12.4. Formation water

In some deep drillings in the quest for either oil or geo-thermal sources, one encounters saline waters beneath the circulating meteoric waters, ranging from brackish to concentrated brines. These are found in different geologic formations and with a large variety of chemical compositions, although usually chloride-dominated. The general term *Formation Water* is used to indicate essentially any stagnant water body found in association with deep formation without implying any generic relationship. As described by Kharraka and Carothers (1986), one can then distinguish between *connate waters*, i.e. waters of marine origin remaining or invading the formations since their genesis, and *metamorphic* and *diagenetic waters* that refer, respectively, to remnants of high-temperature interactions during the metamorphism of rocks or water released from the solid phases during mineral reactions, such as the transformation of gypsum to anhydrate.

Due to the predominance of chloridic waters, it has generally been assumed that most of the formation waters are indeed remnants of marine brines trapped in the marine sediments, or forced to migrate into adjacent formations due to uplift and tectonic activity, where some modification of the chemical character may have occurred by interaction with the new host minerals or by mixing with the prevalent waters in these formations.

The isotopic composition of such waters is expected to provide clues on the evolution of these systems, and indeed, many isotopic measurements on formation waters have been launched worldwide, as reported by Clayton *et al.* (1966), Hitchon and Friedman (1969), Fleischer *et al.* (1977) and Fritz and Frape (1969), among others. As reviewed by Horita (2005), the isotopic composition of marine brines, of continental brines, of *metamorphic* waters or of *diagenetic* waters, are expected to be quite distinct one from another and also from waters of the meteoric cycle. As was discussed in Chapter 11, the marine brines become more enriched in the heavy isotopes of both oxygen and hydrogen than the ocean waters during moderate degrees of enrichment, but at the later stages of evaporation, a reversal of the enrichment will be noticed. The changes of the isotopic composition of the ocean waters over the geologic time span, which are estimated to be $+1‰$ and $-0.5‰$ in $\delta(^{18}O)$, (*vid* Chapter 5) are relatively minor in comparison to the evaporative signature. In contrast, *metamorphic* waters, which are assumed to have evolved at high temperatures in isotopic equilibrium with the rock matrix (there is little isotopic fractionation at these high temperatures), are believed to be close to the composition of the *juvenile waters* as described by Boato (1961), namely slightly enriched in the heavy oxygen isotopes and depleted in Deuterium relative to the ocean waters.

The recorded isotope values of formation waters cover a wide range, indicating varied evolutionary processes. This will be demonstrated, based on a few examples:

- Craig (1966) reported on a brine in the area of the Salton Sea, California, at a depth of 1400 m, reaching temperatures of $>300°C$. At first believed to represent magmatic exhalation, the isotope data shown in Fig. 12.3 clearly indicates a local meteoric origin of the waters, accompanied by increasing Oxygen shift as waters percolate down and pick up salinity from the transected formation. The surface waters in this region, namely Lakes Salton and Mead, as well as the Gulf of California, apparently are not involved in this process.
- Brine samples from the Israeli coastal plain (Fleischer *et al.*, 1977) coming from depths of 1000–2750 meters in an area of oil producing wells, are predominantly chloridic in nature and rather close in chemical composition

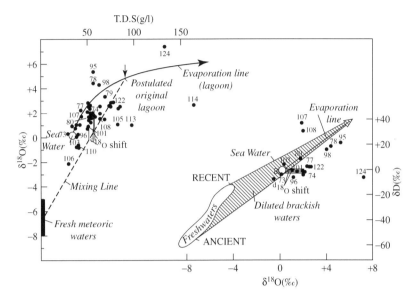

Fig. 12.5.   Isotope composition and salinities of formation waters from deep drillings in the coastal plain of Israel; (note the Oxygen shift of some of the waters).

to seawater, except for some substitution of Mg by Na and Ca ions. Salinities range from that of seawater to about three times this value. As shown in Fig. 12.5 (adapted from Fleischer *et al.*, 1977), in most of the cases, Oxygen-18 is slightly enriched relative to the seawater composition, with little change in $\delta(^2H)$. One prominent exception is sample 124, where evidently, at temperatures above 100°C, complete re-crystallisation of the formation took place accompanied by a large *Oxygen-shift*. Two other outliers, no's 107 and 108, show relatively excessive enrichment in Deuterium.

The marine origin of these brines is evident; it remains ambiguous whether the samples represent a process of evaporative enrichment in lagoonary systems followed by dilution by meteoric waters or a direct interactive process with the host formation, such as oxygen isotope exchange with the carbonates or hydrogen isotope exchange with clays. However, sample numbers 106 and 110 show evidence of the admixture of meteoric waters.

- Oil field brines from the Illinois, Michigan and Alberta sedimentary basins (Clayton *et al.*, 1966): These continental brines evidently are derived from meteoric waters that match the local isotopic composition or from local paleowaters. The oxygen isotope enrichment relative to the meteoric waters, which was found to be strongly correlated with the temperature, is explained by these authors as being the result of an isotopic exchange with the carbonate rocks. However a similar set of data from deep drillings in the Jordan Rift Valley (Fleischer *et al.*, 1977) were there interpreted as deriving from waters of "Inland Brine Lakes" that mix with the local meteoric waters.

Most of these data are situated on the *δ-plot* to the right of the MWL. However, Fritz and Frape (1982) and Kloppmann *et al.* (2002) reported for brines from the Canadian shield isotope values situated on the other side of the MWL, which presumably originated from *diagenetic* rather than *connate* waters.

There is, as of now, no consensus on all the processes at play in forming these formation waters. Horita (2005) summarised the situation as follows: "The origins and evolution of sedimentary brines still remain elusive even after almost four decades of active research." Table 12.1 summarises the possible origins of the formation waters as well as the transformations in chemical and isotopic character through interactions with the host formation.

Table 12.1. The evolution of formation waters.

12.1.A	Water Origin	Chemical Character	Isotope signature
1.	Connate ocean water	Seawater chemistry	MOW $\pm$ 1‰ in δ (^{18}O)
2.	Marine lagoon	Concentrated seawater	Enriched OW in both ^{2}H and ^{18}O along EL
3.	Continental Brine lake	Na–Ca–Mg	Enriched Meteoric water along EL
4.	Magmatic	varied (function of formation)	$\approx$ juvenile water
5.	Diagenetic	Ca-sulphatic	above the GMWL

12.1.B	Diagenetic Process	Isotope Signature
a)	Evaporation	Evaporation line
b)	Mixing with local water	Along mixing line
c)	Ultrafiltration	
d)	Interaction with Carbonate rocks	Oxygen shift
e)	Interaction with clay minerals	Hydrogen shift
f)	Gypsum precipitation	
g)	Geo-thermal venting	

Chapter 13

The Continental Scale Water Balance and Its Isotopic Signature

As was described in the previous chapters, both the surface and sub-surface drainage fluxes acquire distinct isotopic signatures at the land-surface interface, that depend on the precipitation input, the local water balance and eco-hydrological structure. As the surface and ground waters, respectively, then drain downstream over an increasing geographic extent, this initial isotope composition is modified. Foremost factor for the modification of the isotopic composition during the passage downstream are the additional local inputs by either tributaries to the surface water flow or additional recharge to the groundwater as long as it is still an unconfined aquifer system. In addition, secondary changes occur, by the interaction of the water fluxes with their environments. In the case of the groundwater flux, such secondary interactions are relatively minor, consisting of a possible isotope exchange (especially of the Oxygen isotopes) with the aquifer formation and mixing with stationary or deeper groundwaters. In the case of surface water, the major secondary factor of change of the isotope composition as the waters discharge downstream is that associated with evaporative water loss. Other factors to reckon with are the interchange with natural and human consumers, as elaborated in the IRCM scheme discussed in Section 9.3.

A primary distinction has to be made between the situation under humid and dry climates (Gat and Airey, 2006). In humid and temperate climates, where the precipitation inputs exceed the evapo/transpiration losses, the drainage basins are of an *exorheic* nature with a potential discharge of both surface and subsurface streams to the ocean. An increasing portion of the drainage downstream occurs by surface flow which is augmented by spring discharges of the groundwaters along the flow-path. In contrast, under dry conditions, the drainage basins can be of an *endorheic* nature, where the surface streams empty into an evaporative water body such as a *terminal lake*, so that any residual discharge to the ocean will be by groundwater only.

In order to assess the isotope composition of the continental discharge fluxes in comparison to the ones as given by the local ITF's, the detailed meteorological pattern, the surface morphology, land-use and eco-hydrological makeup are all factors to consider. As the geographic extent increases while both surface and groundwater drain downstream, a number of additional links between them are activated, as shown schematically in Fig. 13.1. Spring and interflow discharges augment the surface flow while, on the other hand, additional groundwater recharge from the surface waters can occur through leakage from downstream impoundments or by bank infiltration. Since the surface waters are continuously exposed to evaporation, they accumulate more and more of the "evaporative isotopic signature" that can then also be transmitted into the groundwater flow by the additional recharge link. This is especially significant under more arid conditions (c.f. Fontes and Gonfiantini, 1967). Under humid conditions, on the other hand, since the groundwaters are "shielded" from direct exposure to evaporation, the addition of the spring discharge usually serves to dilute and counter the evaporative signature of the surface waters.

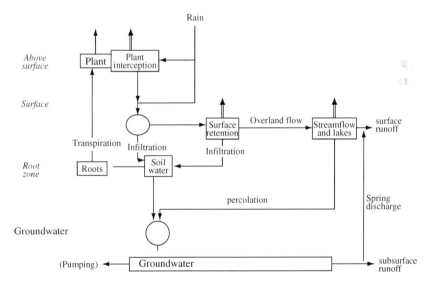

Fig. 13.1. Schematic representation of the coupling between surface and subsurface flows on a continental scale.

During a continental passage, one distinguishes between the wind-side and lee-side of a range of elevated ground that serves as a continental water-divide. On the wind-side of the continent, the runoff runs counter-current to the air stream and on the average, drains the integrated amount of precipitation (corrected for evapo-transpiration losses), often however, in a delayed fashion due to a seasonal or longer holdup in large lakes and aquifers, especially in the case of solid precipitation elements. Under an ideal Rayleigh-rainout scenario, the atmospheric moisture is depleted in the heavy isotope content commensurate with the loss of moisture by precipitation, following more or less a "Meteoric Water Line", thus retaining a constant value of the *d-excess* parameter. If evapo-transpiration is a significant portion of the water balance, then there is an apparent reduction in the degree of the Rayleigh rainout effect on the isotopic composition of the atmospheric moisture. To the extent that the E-T flux returns the incoming water in an unfractioned manner, as is the case for transpiration and a terminal system as well as when a surface water body dries up completely, it can be measured by comparison of the actual change in isotope composition of the atmospheric waters to the change expected based on the Rayleigh relation for the amount of precipitation, as was successfully applied for an estimate of the water balance of the Amazonian rain forest (Salati *et al.*, 1979). In Box 13.1, data from the Amazon basin is elaborated. In the absence of sufficient data on the isotope composition of the atmospheric moisture, the isotope composition of the precipitation is usually taken as a proxy on the assumption of isotopic equilibrium between precipitation and vapour, as was discussed in Section 6.2.

Partial evaporation from an open water surface or from within the soil return a fractioned flux in which the evaporated waters are depleted in the heavy isotopes of both hydrogen and oxygen following an Evaporation Line rather than the MWL, as was discussed in Chapter 4. Under such circumstances, the isotopic composition of the atmospheric waters is not totally restored and moreover, is moved away from the original MWL. As shown for the example of the Great Lakes of North America (Gat *et al.*, 1994) the degree of recycling of moisture by this process can then be quantified by recording the change in the *d-excess* parameter of the downwind atmospheric waters, as described in Box 13.2.

In a more arid environment with its *endorheic* nature, more and more of the runoff is lost by evaporation as the scale of the basin increases, often terminating in highly saline terminal systems. In contrast to the situation described above, the increase of the *d-excess* of the atmospheric moisture

Box 13.1. Water balances in the Amazon Basin, based on stable isotope data.

Accounting for the continental water balance simply by subtracting the sum of the precipitation from the atmospheric moisture is misleading in those areas where evapo-transpiration is a significant factor in the atmospheric water balance. In the Amazon Basin, as an example, this becomes apparent when the isotope composition of precipitation in the inner part of the basin is compared to the one expected based on the "Rayleigh Rainout" relationship, as shown in Figure A.

A basin-wide balance would require, however, precipitation data from throughout the basin, as well as down-wind from it.

The same purpose can be achieved by means of a regional runoff model based on isotope data of both the atmospheric influx and the river runoff, on the assumption that the runoff presents the integrated amount of precipitation corrected for the re-evaporation component and that the re-evaporated moisture's isotope composition is unfractionated relative to the precipitation input, as is the case for the transpiration component. The part of the return flux due to evaporation is treated in this simple model the same way.

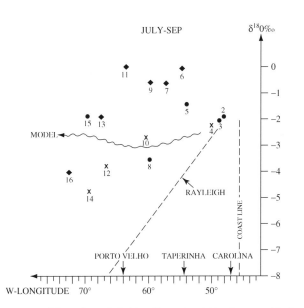

Fig. A. Isotope composition of precipitation throughout the Amazon Basin, based on Salati *et al.*, 1979. The lower line represents the value that would be expected for a Rayleigh-rainout model and the upper line is based on a model incorporating the re-cycled water. Data points are from the GNIP network. **x**, • and ♦ stand for the northern, central and southern part of the basin, respectively.

(*Continued*)

Box 13.1. (*Continued*)

The water balance equation for the watershed is given by Equation 1, where F_{in} and F_{out} are the vapour influx and downwind flux, respectively, and Ru the runoff discharge.

$$F_{in} = Ru + F_{out} \tag{1}$$

The isotope equation (Eq. 2) is then as follows, where δ_{in}, δ_{out}, δ_{Ru} are the isotopic values for these fluxes

$$\delta_{in} \cdot F_{in} = \delta_{Ru} \cdot Ru + \delta_{out} \cdot F_{out} \tag{2}$$

Defining the water yield of the basin as $f = F_{out}/F_{in}$ (defined thus, f is corrected for the recycled moisture) then the following expression for f results based on the isotopic data:

$$f = (\delta_{in} - \delta_{Ru})/(\delta_{out} - \delta_{Ru}) \tag{3}$$

Provided the values of the isotopic composition of the runoff (δ_{Ru}) and of the vapour influx (δ_{in}) are known by measurement [the latter can be approximated by measurement of the precipitation ($\delta_{p,0}$) on the assumption of precipitation being formed under equilibrium conditions so that $\delta_p = \delta_{in} + \varepsilon^*$], then Eq. 3 remains with 2 unknown quantities, namely f and δ_{out}. These can be evaluated in 2 ways: either by assuming a Rayleigh Rainout Law throughout the basin, so that:

$$\delta_{out} = \delta_f = \delta_{in} + \varepsilon^* \cdot \log f \tag{3a}$$

this value can then be introduced into Eq. 3. Alternatively it can be assumed that the isotopic composition of the runoff represents the averaged value of precipitation over the whole basin, approximated by: $\delta_{Ru} = (\delta_0 + \delta_f)_p/2$.
In that case:

$$f = (\delta_{Ru} - \delta_{p,0} + \varepsilon^*)/(\delta_{p,0} - \delta_{Ru} + \varepsilon^*) \tag{3b}$$

the values of f found by either of these solutions differ just a little. Obviously the test of this procedure would be given by measurement of the downwind value of δ_{out}.

Based on monthly and three-monthly averaged values of the Amazon runoff near the mouth (at Obidos) and of the precipitation at Belem, values for f were found to change seasonally during the period from 1982 to 1984. Higher values of f were calculated for the internal drainage basins of Rio Solimoes and Rio Negro, based on measurements at Manaus. The full data set is available in Tables 1 and 2 in Gat (2000).

Increased possibilities open up with the direct measurement of the isotope composition of vapour by observations aloft.

Box 13.2. The estimate of the contribution of evaporation from a lake to the atmospheric moisture, based on the change in the *d-excess* parameter.

In order to evaluate the change of the d-excess value of the atmospheric water's isotope composition resulting from the admixture of evaporated waters, one needs to compare the d value of the evaporate (d_E) with that of the ambient moisture (d_A) into which it is introduced. Based on Eq. 4.9a where the δ value of the evaporate is given for either the oxygen and hydrogen isotopes, and remembering the definition of $d = \delta(^2H) - 8 \cdot \delta(^{18}O)$ one obtains the following value of d_E:

$$\delta_E \approx [\delta_L - h \cdot \delta_A - \varepsilon]/(1-h); \quad \varepsilon = \varepsilon^* + \Delta\varepsilon; \quad \Delta\varepsilon = (1-h) \cdot \theta \cdot n \cdot C_k$$

$$d_E = [d_L - h \cdot d_A - (^2\varepsilon \cdot^* - 8 \cdot {}^{18}\varepsilon^*) - (1-h) \cdot \theta \cdot \{^2C_k - 8 \cdot {}^{18}C_k\}]/(1-h)$$

where obviously the values of $^2\varepsilon^*$ and ^2C_k refer to the heavy hydrogen isotope and the equivalent terms labeled with 18 refer to the Oxygen-18 substituted molecules.

The term $(^2\varepsilon \cdot^* - 8 \cdot {}^{18}\varepsilon^*)$ is close to the value of zero at ambient temperatures while $\{^2C_k - 8 \cdot {}^{18}C_k\} = -107\permil$. Thus in permil notation:

$$(d_E - d_A) = (d_L - d_A)/(1-h) + 107 \cdot \theta$$

This value, i.e. $(d_E - d_A)$, is then a scaling factor against which to assess the contribution of the evaporated moisture to the atmospheric water balance. Unlike the value of δ_A which is hard to establish, the value of d_A can be given (based on the local precipitation) with considerable assurance. Indeed the only factor in doubt is the value of θ.

Based on such an approach, the increase of $3.5\permil$ in the *d-excess* of atmospheric moisture over the Great Lakes region of north America during summer was estimated to account for a contribution of between 5.7% to 9.5% of moisture evaporated from the lakes [Gat *et al.*, 1994].

---

diminishes as a larger and larger fraction of the water is evaporated. The distinction between the evaporation and transpiration fluxes based on the atmospheric water balance and its isotopic composition, which is a valuable diagnostic tool under more humid and semi-arid conditions, cannot then be applied. However, the comparison between the expected isotope depletion in the air moisture due to the Rayleigh rainout effect and the measured value can then be applied again in a straightforward manner as a measure of the total amount of evapo-transpiration. Such endorheic systems are typically more prevalent on the leeside of the continental divide and in interior closed basins such as, for example, within the Rocky Mountains and the Jordan Valley.

The Amazon basin is a rare case where the ideal countercurrent scenario is applicable. In other cases, the isotope balance on a continental scale is rather difficult to establish, in view of varying vapour sources over the catchment, the fragmentary nature of the runoff data and the insufficient data on the isotopic composition of water vapour on the leeside of the continents. The isotope-aided atmospheric models, to be soon discussed, then come closest to enabling the assessment of the regional water balances.

13.1. Water isotopes on the continental lee-side

During the passage over the continents the residual atmospheric moisture, as discussed above, accumulates re-evaporated vapour with its d-excessive signature. As the air-stream then approaches the ocean on the leeside of the continent (usually relatively cold and with a low vapour content), the rapid reloading of the air moisture over the sea also results, in the first instance, in an increase of the d-excess of the atmospheric moisture (*vid.* Chapter 5). Under conditions of re-entry of such air-masses onto the coastal lands, this may then lead to an erroneous interpretation of the continental water balance.

Prominent examples are the situation on the Atlantic seacoast of the north American continent, the China sea area on the leeside of the Asian mainland and the Mediterranean coastal lands in southern Europe.

13.2. Isotope aided GCMs

During the mid-eighties, a new modeling tool was developed, namely Atmospheric General Circulation Models fitted with water isotope diagnostics; these were, in order of their formulation, the LMD in Paris (Joussaume *et al.*, 1984), the GISS (Jouzel *et al.*, 1987), the ECHAM model (Hoffmann *et al.*, 1998) and a variety of more recent ones. The purpose was, on the one hand, to provide a calibration tool for the model assumptions by comparing the model output with the measured GNIP data set. On the other hand, it was expected that these models could serve as interpolation tools to augment the very scattered GNIP data set on a wider geographic basis. Further, it was intended that these models would be useful in the paleoclimate research, enabling a calibration of the proxy isotope data in terms of climatic parameters such as temperature and amount of precipitation. Being based on a rather coarse geographic grid scale and on the Craig-Gordon Evaporation model and the Rayleigh rainout concept, they perform best over the oceanic and large continental areas. As was summarised by Hoffmann *et al.* (2005): "Isotopic AGCMs are able to simulate reasonably well the seasonal cycle and spatial distribution of water isotopes." However, on an inter-annual time scale, the model does not perform as well, and especially so in regions with little rain or in the region of the ITCZ where a Rayleigh Rainout Law does not apply. The more continuous information on the isotopic composition of the atmospheric moisture (even in the absence of precipitation) as provided by the laser spectroscopic measurements will be of great value in improving the model performance. Further, a better control on the isotope fractionation during evapo-transpiration at the land surface for varied local conditions is needed. A new initiative of improving land-surface parametrization schemes using stable isotopes (iPILPS) as described by Henderson-Sellers (2006), aims at providing a more realistic input into the regional models.

Chapter 14

Isotopes and Climate Change

14.1. Oxygen and Hydrogen isotope composition of proxy materials as tools in paleo-climate research

Since the isotope composition of meteoric water is correlated with climatic and hydrologic parameters, such as the temperature dependence of the isotope content of precipitation, the d-excess value as a measure of the humidity deficit at the site of evaporation and the evaporative enrichment of the stable isotopes in surface waters, as a measure of the water balance in lakes, it seems reasonable to attempt to use the isotope composition of waters unrelated to the present water cycle as tools for reconstructing past climate scenarios. However, water originating from past periods is conserved and met with only under special circumstances, among them:

- ice accumulation in glaciers,
- paleo-waters, i.e. meteoric groundwater recharged during past climate periods,
- fluid inclusions in mineral accumulations,
- interstitial waters in sediments.

Such water archives are alas often restricted to the more recent past. However, a variety of more widespread materials related to the water cycle has been utilized as climate proxies, by attempting to infer the isotopic composition of the waters from that of the proxy, based on the recognition of the rules of partitioning of the isotopes between the material concerned and the water environment. Table 14.1 lists some of the materials employed. We note that only in a few cases can one derive information on both the oxygen and hydrogen isotopes, thus losing the possibility of recording the d-excess value of the water. On the other hand, additional parameters

Table 14.1. Isotopic climate proxies

Analysed material	Isotopes	Environmental parameters
Ice Cores[a]:		
Water	^{18}O ^2H	Temperature, Atmospheric Water Balance
Lake sediments[b]:		
Carbonates	^{18}O ^{13}C ^{14}C	Watershed Water Balance
Silicates	^{18}O	Temperature
Aquatic Cellulose	^{18}O	
Ocean sediments[c]:		
Carbonates	^{18}O ^{13}C	
Silicates	^{18}O	Temperature
Phosphates	^{18}O	
Plant remains[d]:		
Cellulose	^{18}O ^2H ^{13}C ^{14}C	IC of Soil Water Humidity
Speleothems[e]:		
Carbonates	^{18}O ^{13}C ^{14}C	IC Precipitation/Recharge
Fluid inclusions	^{18}O ^2H	
Snails (shells)[f]:		
Carbonate	^{18}O $^{13}C^{14}C$	Aridity
Mineral deposits:		
Hydration water[g]	^{18}O ^2H	IC of ambient waters
Clays[h]	^{18}O ^2H	""
Palaeo-Groundwater[j]:		
Water	^{18}O; ^2H; ^3H; ^{14}C	Watershed Water Balance

[a] Dansgaard *et al.* (1969); Arnason (1981) ;Thompson and Davis (2007);
[b] Stuiver (1970) ; Gasse (2007)
[c] Epstein *et al.* (1951, 1953); Craig (1965)
[d] Epstein, Thompson and Yapp (1977); Switsur and Waterhouse (1998); Gat *et al.* (2007)
[e] Schwarcz (1986); Bar-Matthews *et al.* (1997)
[f] Yapp (1979); Magaritz & Heller (1980); Goodfriend *et al.* (1989)
[g] Fontes and Gonfiantini (1967); Sofer (1978)
[h] Savin and Epstein (1970)
[j] Edmunds (2007).

relevant to the formation of the proxy, such as temperature, can be determined. Further, the dating of some of these materials is more reliable than that of the paleowater.

The different proxies vary in the time resolution of the recorded information, ranging from the annual signals discernable in tree rings and some ice cores, to medium scales in lake deposits (depending on the sedimentation rates) to rather low resolution of ±1000 years in large groundwater bodies (Edmunds, 2007).

It has to be emphasised that the full interpretation of the isotope record in the proxy archives is dependant on reliable dating of the material. Further additional geochemical and ecological evidence provided by trapped or dissolved gases, salt accumulations, micro-fossil analyses, etc., can provide auxiliary information.

Finally, it has to be recognised that post-depositional changes and interactions may obscure the situation somewhat and have to be taken into account.

Some relevant references are given in the footnotes to Table 14.1.

14.1.1. *Glaciers*

Ice accumulation on glaciers, Arctic, Antarctic or at high altitude, is the most tell-tale climate proxy, and deep ice-cores have been extensively researched. Their advantage in this respect is that the precipitation itself is the material that accumulates in a sequential fashion, enabling us to record both oxygen and hydrogen isotopes. The relatively large accumulation rate enables an annual time resolution, taking advantage of the seasonal variability of the isotope composition of the precipitation. However, this ice record is obviously limited to the cold regions and also in the recorded time span. On a deep ice core on Greenland (Camp Century), the record goes back to 100,000 years (Dansgaard *et al.*, 1969) and was extended to close to 150,000 years in subsequent drillings. In the low latitude, high altitude ice the record is shorter but throughout the Holocene, the variations mirror those of the polar data indicating a global control on the isotope content.

An additional advantage of the ice core record is that other environmental indicators such as dust, pollen and gas inclusions (especially CO_2) are also frozen in and can serve as additional evidence for the climatic condition at that time. On the debit side, however, is the fact that the stage of formation of the ice from the precipitation elements is accompanied by changes in the isotope composition. As discussed in chapter 7.2, these changes themselves are climate-dependent, especially on the thermal diurnal and seasonal cycle. Therefore, caution is advised in interpreting the variations strictly in terms of the ambient temperature based on the present-day isotope-temperature relationship for precipitation.

The isotopic ice record has been extensively reviewed over the years, with Arnason (1981), Moser and Stichler (1980), Thompson and Davis (2007) being some notable examples. A delightful personal account of this work has been published recently by W.Dansgaard (2004).

14.1.2. *Ocean and lake sediments*

The basic premise for the use of ocean and lake sediments as a climate proxy is that they are deposited with their oxygen isotopes in isotopic equilibrium between the water and the sedimenting mineral. This equilibrium fractionation factor has a marked temperature dependence, as expressed in Equation (3.2). For the case of a carbonate mineral (calcite), the following values were given (Lerman and Clauer, 2007):

$$\delta^{18}O_{calcite} - \delta^{18}O_{water} = 2780/T^2 - 0.00289$$

In the range of ambient temperatures, this corresponds to a change of about $\pm 1\%_0$ in the value of $\delta^{18}O_{calcite}$ for a change of $\pm 5°C$, assuming equilibrium with water with a constant $\delta^{18}O$ value. Considering the analytical error margin, this would enable a temperature definition of up to $\pm 0.5°C$.

The constancy of the isotope composition of the water is more easily maintained in the ocean, because of the relatively small changes in the isotopic composition of ocean waters. Even the uncertainty of the glacial increment (*vid.* chapter 5) is encompassed within this limit of sensitivity of $\pm 5°C$.

The method can be expanded to other situations, such as sediments from coastal waters or inland lakes, provided an independent assessment of the isotope composition of the water can be made. An obvious possibility is the availability of trapped interstitial waters within the sediments. As summarised by Lister (quoted in Gat and Lister, 1995):

> "The isotopic record in the interstitial waters has the advantage over that in the carbonate sediments, in that it can provide a record of the ratios for the water isotopes of both oxygen and hydrogen, so that even in the case where the oxygen isotope composition is changed by diagenetic interactions, the hydrogen isotopes survive to tell the story. Further temperature variations in the lake obviously are not a factor in this case. On the other hand, the pore water system cannot be regarded to be a closed one for any length of time so that the memory of the interstitial waters is limited to a few thousand years at most (Stiller *et al.*, 1983)."

Another possibility for obtaining a temperature record in the waters, first suggested by Urey as reported by Craig (1965), is the co-existence of an additional sediment with a different temperature-isotope composition scale. Such sediments could be phosphorites or siliceous minerals. However, as first reviewed by Longinelli and Nuti (1965), the phosphorus minerals show a very similar temperature curve to the carbonates, so that a very high precision in measurements is indicated. In the case of limnic sediments, where a large sedimentation rate may enable a very detailed time resolution, the method may enable us to chart a seasonal temperature cycle without the necessity of spelling out the water's isotope composition exactly. In the case of limnic sediments, the changes of the water composition over long periods can be much larger than the effect of any foreseeable temperature change, so that in such a case, these changes in the water balance of a lake can be inferred from the profile of the isotopic composition in the sediment layers.

In addition, the following limitations of the method are recognised, namely, alterations by diagenetic processes such as recrystallisation and interaction with other waters such as meteoric water, and the so-called "vital effects" in bio-mediated sedimentation which result in some of the sediments not being formed in isotopic equilibrium with the water.

The subject has been extensively reviewed over the years. In addition to the cited reviews of Craig (1965) and Lerman and Clauer (2007), the reviews by Duplessy (1979) and Gasse (2005) are noteworthy.

Reconstructing both the hydrogen and isotope composition of a lake's water in relation to the Meteoric Water Line has the advantage that it might enable one to distinguish between the effect of evaporation and the changes in the isotope composition of the inflow. The limitations of using interstitial waters in that respect were mentioned above. Another possibility is given if the hydrogen isotopic signal of hydrogen is preserved in sediments such as caolinite, illite etc. along with the carbonates or, even preferably, in cellulose of aquatic plants preserved within the sediments.

14.2. Effect of climate change on the isotope signature in the hydrologic cycle

Global climate change will affect the total hydrologic cycle, from the oceanic source region, followed by the atmospheric pathways and their water balance, the processes at the land/ecosphere/atmosphere interface that determine the recharge flux and/or the surface runoff and finally, the continental water balance. In almost all of these phases, a change in the isotopic signature is to be expected.

The isotopic composition of the ocean waters is expected to change mainly as a result of the melting of the Arctic and Antarctic ice caps. This *"glacial increment"* was estimated to be of the order of $-0.5‰$ to $+1‰$ in $\delta^{18}O$ around the value of SMOW, for the full circle of glacial and interglacial epochs. Differences in the slight enriched values of the surface waters, resulting from the evaporation, are expected to be within this range. However, in coastal regions and, in particular, in semi-enclosed Mediterranean seas such as the Baltic, the Black Sea or the Persian Gulf, where the river discharge is a major factor in their water balance, the variations in the isotope composition of the seawater can be appreciable (Anati and Gat, 1989).

As is evident from Eq. (4.10) of the Craig-Gordon Evaporation model and the companion atmospheric mixing model, shown in Fig. 5.5, the changes in the moisture regime of the atmosphere at the sites of evaporation are of even greater consequence. Since such changes (and in particular the moisture content of the atmospheric boundary layer) also determine the d-excess value of the evaporation flux, which is later essentially conserved during the rainout process, this parameter has a special significance in the paleo-climatic reconstruction. As was discussed in section 12.2, there is a consistently reduced d-excess value of paleowaters related to the last glacial period, that is interpreted as a higher moisture content over the oceans at that time (Merlivat and Jouzel, 1979).

In the atmospheric part of the water cycle, the most immediately discernable effect of the changing climate is related to the temperature change as it affects the precipitation pattern and thus the Rayleigh rainout pattern. However, one must take care in interpreting the changing isotope composition over past periods simply in terms of the present-day relationship between the isotope composition of the precipitation and temperature, as given for example by Gourcy *et al.* (2005) (*vid.* section 6.1.2). Not only is there a change in the source characteristics of the marine moisture, but the air trajectories and their precipitation patterns can also be affected, especially in climate transition zones such as the Mediterranean Sea region as presented for example by Gat and Carmi (1987).

As the precipitation then falls on the ground, one expects great changes in the ITF (*vid.* section 8.1) for the transition to either infiltration into the ground, surface runoff or evapo-transpiration under a climate change scenario. Even if one assumes no major changes in the surface structure and land-use pattern, any change in the precipitation pattern and especially its seasonal distribution as well as changes of temperature, humidity and wind regime will affect the partitioning of the precipitation at the Land/Biosphere/Atmosphere interface and thus of its isotopic signature. These changes will undoubtedly be intensified by ecological changes resulting from the climate change, especially of the plant cover.

One major factor in changes of the hydrological regime and with it of the isotopic composition is to be expected in case of snow replacing fluid precipitation (and vice versa). A notable change is introduced by the time delay in the transition through the interface until melting sets in, as well as a possible shift in the distribution between the effluent fluxes.

On the watershed scale, the changes in the evaporative water loss along the riverflow (in particular in overflow wetland areas and from lakes) are a function of the climate. In view of the great sensitivity of the isotope composition to the evaporative regime, the isotope signature will respond notably to such a change and can serve as a monitor to this. However, as discussed by Gat and Lister (1995), the isotope effect imposed by changes in the terrain and vegetation cover in the catchment source regions can, by itself, introduce an isotopic signal change which was estimated by these authors as amounting to several permil units in $\delta^{18}O$.

References

Adar, E. M., Gev, I., Lipp, J., Yakir, D. and Gat, J. R. (1995). Utilisation of oxygen-18 and deuterium in stem flow for the identification of transpiration sources: Soilwater versus groundwater in sand dune terrain. *Application of Tracers in Arid Zone Hydrology* (E. Adar and C. Leibundgut, eds.) IAHS Publ. 232: 329–338.

Allison, G. B. (1982). The relationship between ^{18}O and Deuterium in water in sand columns undergoing evaporation. *J. Hydrology* 55: 163–169.

Allison, G. B. and Leaney, F. W. J. (1982). Estimation of isotopic exchange parameters using constant-feed pans. *J. Hydrology* 55: 151–161.

Allison, G. B., Barnes, C. J. and Hughes, M. W. (1983). The distribution of Deuterium and ^{18}O in dry soils, experimental. *J. Hydrology* 64: 377–397.

Allison, G. B., Gat, J. R. and Leaney, F. W. J. (1985). The relationship between Deuterium and Oxygen-18 delta values in leaf water. *Chemical Geology (Isotope Geoscience Section)* 58: 145–156.

Anati, D. A. and Gat, J. R. (1989). Restricted marine basins and marginal sea environments. Chapter 2 in *Handbook of Environmental Isotope Geochemistry* (P. Fritz and J.-Ch. Fontes, eds.), 3: 29–74.

Aravena, R., Suzuki, O. and Pollastri, A. (1989). Coastal fog and its relation to groundwater in the IV region of northern Chile. *Chemical Geology (Isotope Geochemistry Section)* 79: 83–91.

Arnason, B. (1969). The exchange of hydrogen isotopes between ice and water in temperate glaciers. *Earth Planet. Science Lett.* 6: 423–430.

Arnason, B. (1981). Ice and snow hydrology. *Stable Isotope Hydrology: Deuterium and Oxygen-18 in the Water Cycle* (J. R. Gat and R. Gonfiantini, eds.), Technical Report Series No. 210, IAEA, Vienna, Chapter 7: 143–175.

Ayalon, A. (1998). Rainfall-recharge relationships within a karstic terrain in the eastern Mediterranean semi-arid region, Israel: $\delta^{18}O$ and δD characteristics. *J. Hydrology* 297: 18–31.

Bar-Matthews, M., Ayalon, A. and Kaufman, A. (1997). Late Quarternary pale-oclimate in the Eastern Mediterranean region from stable isotope analysis of speleothems in Soreq Cave, Israel. *Quatern. Res.* 47: 155–168.

Barkan, E. and Luz, B. (2005). High Precision measurement of $^{17}O/^{16}O$ and $^{18}O/^{16}O$ ratios in H_2O. *Rapid Comm. Mass Spectrometry* 19: 3737–3742.

Barkan, E. and Luz, B. (2007). Diffusivity fractionation of $H_2^{16}O/H_2^{17}O$ and $H_2^{16}O/H_2^{18}O$ in air and their implications for isotope hydrology. *Rapid Comm. Mass Spectrometry* 21: 2999–3005.

Barnes, C. J. and Allison, G. B. (1983). The distribution of Deuterium and ^{18}O in dry soils, 1. Theory. *J. Hydrology* 60: 141–156.

Barnes, C. J. and Allison, G. B. (1988). Tracing of water movement in the unsaturated zone using stable isotopes of Hydrogen and Oxygen. *J. Hydrology* 100: 143–176.

Beyerle, U., Purtschert, R., Aeschbach-Hertig, W., Imboden, D. M., Loosli, H. H., Wieler, R. and Kipfer, R. (1998). Climate and groundwater recharge during the last glaciation in an ice-covered region. *Science* 282: 731–734.

Baily, I. H., Hulston, J. R., Macklin, W. C. and Stewart, J. R. (1969). On the isotopic composition of hailstones. *J. Atmosph. Sciences* 26: 689–694.

Bigeleisen, J. (1962). Correlation of tritium and deuterium isotope effects. *Tritium in the Physical and Biological Sciences*, 1, IAEA, Vienna, pp. 161–168.

Bleeker, W., Dansgaard, W. and Lablans, W. N. (1966). Some remarks on simultaneous measurements of particulate contaminants including radioactivity and isotopic composition of precipitation. *Tellus* 18: 773–785.

Boato, G. (1961). Isotope fractionation processes in nature. *Summer Course on Nuclear Geology, Varenna*, Lab.di Geolog. Nucleare, Pisa, pp. 129–149.

Bolin B. (1959). On the use of tritium as a tracer for water in nature. *Proc. 2nd Conf. on the Peaceful Uses of Atomic Energy*, Geneva, UN. 18: 336–344.

Bowser, C. J. and Gat, J. R. (1995). On the process of Lake-Ice formation. *International Symposium on Isotopes in Water Resources Research* IAEA-SM-336/44P: 209–210.

Bricout, J., Fontes, J. Ch. and Merlivat, L. (1972). Sur la composition en isotopes stable de l'eau des jus d'orange. *C. R. Acad. Sci., Paris* Ser. D, 274: 1803–1806.

Brinkmann, R., Eichler, R., Ehhalt, D. and Muennich, K. O. (1963). Ueber den Deuterium-gehalt von Niederschlags- und Grundwasser (1963). *Naturwissenschaften* 50: 611–612.

Broderson, Ch., Pohl, St., Lindenlaub, M., Leibundgut, Ch. and v. Wilpert, K. (2000). Influence of vegetation structure on isotope content of throughfall and soil water. *Hydrological Processes* 14: 1439–1448.

Brutsaert, W. (1965). A model for evaporation as a molecular diffusion process into a turbulent atmosphere. *J. Geophys. Res.* 70: 5017–5024.

Cappa, C. D., Hendricks, M. B., DePaolo, D. J. and Cohen, R. C. (2003). Isotopic fractionation of water during evaporation. *J. Geophys. Res.* (Atmosph.) 108: 4524–453.

Chapman, S. and Couling, T. G. (1951). Mathematical theory of non-uniform gases. Cambridge University Press. Chapter 10.

Chow, V. T. (1964). *Handbook of Applied Hydrology*. McGraw Hill, New York.

Clayton, R. N., Friedman, I., Graf, D. L., Mayeda, T., Meents, W. F. and Shimp, N. F. (1966). The origin of saline formation waters — I. Isotopic composition. *J. Geophys. Res.* 71: 3869–3882.

Clayton, R. N. (1993). Oxygen isotopes in meteorites. *Ann. Rev. Earth Planet. Science* 21: 115–149.

Cook, P. G., Edmunds, W. M. and Gaye, C. B. (1992). Estimating Paleo-recharge and Paleoclimate from unsaturated zone profiles. *Water Resources Res.* 28: 2721–2731.

Cooper, L. W. and DeNiro, M. J. (1989). Covariance of oxygen and hydrogen isotopic composition in plant water: Species effect. *Ecology* 70: 1619–1628.

Coplen, T. B. and Hanshaw, B. B. (1973). Ultrafiltration by a compacted clay membrane. *Geochim. Cosmochim. Acta* 17: 2295–2310.

Coplen, T. B., Hopple, J. A., Bohlke, J. K., Peiser, H. S., Rieder, S. E., Krouse, H. R., Rosman, K. J. R., Ding, T., Vocke, R. D. J., Revesz, K. M., Lamberty, A., Taylor, P. and DeBievre, P. (2002). *Compilation of Minimum and Maximum Isotope Ratios of Selected Elements in Naturally Occurring Terrestrial Materials and Reagents*. U.S Geological Survey, Reston, Va.

Craig, H., Boato, G. and White, D. E. (1956). Isotope geochemistry of thermal waters. *Proc. 2nd Conference on Nuclear Processes in Geologic Settings*. National Research Council, 400: 29–38.

Craig, H. (1961a). Isotopic variations in meteoric waters. *Science* 133: 1702–1708.

Craig, H. (1961b). Standards for reporting concentrations of deuterium and oxygen-18 in natural waters. *Science* 133: 1833–1834.

Craig, H., Gordon, L. I., Horibe, Y. (1963). Isotope exchange effects in the evaporation of water, 1. Low temperature experimental results. *J. Geophys. Res.* 68: 5079–5087.

Craig, H. (1963). The isotope geochemistry of water and carbon in geothermal areas. *Nuclear Geology in Geothermal Areas* (E. Tongiorgi, ed.) Lab. di Geologia Nucleare, Pisa. pp. 17–53.

Craig, H. (1965). The measurement of oxygen isotope paleotemperatures. *Stable Isotopes in Oceanographic Studies and Paleotemperature* (E. Tongiorgi, ed.) Lab. di Geologia Nucleare, Pisa. pp. 161–182.

Craig, H. and Gordon, L. I. (1965). Deuterium and oxygen-18 variations in the ocean and the marine atmosphere. *Stable Isotopes in Oceanographic Studies and Paleotemperature* (E. Tongiorgi, edtr) Lab. di Geologia Nucleare, Pisa. pp. 9–130.

Craig, H. (1966). Isotopic composition and origin of the Red Sea and Salton Sea geothermal brines. *Science* 154: 1544–1548.

Cuntz, M., Ogee, J., Farquhar, G., Peylin, Ph. and Cernusak, L. A. (2007). Modelling advection and diffusion of water isotopologues in leaves. *Plant, Cell and Environment* 30: 892–909.

D'Amore, F., Gerardo-Abaya, J. and Arnorsson, St. (2002). Hydrogen and Oxygen isotope composition during boiling. *Isotopic and Chemical Techniques in Geothermal Exploration, Development and Use* (St. Arnorsson, ed.) IAEA, Vienna, pp. 229–240.

Dansgaard, W. (1954). The O-18 abundance in fresh water. *Geochim. Cosmochim. Acta* 6: 241–260.

Dansgaard, W. (1960). The content of heavy oxygen isotopes in the water masses of the Philippine Trench. *Deep Sea Res.* 6: 346–350

Dansgaard, W. (1964). Stable isotopes in precipitation. *Tellus* 16: 436–468.

Dansgaard, W., Johnsen, S. J., Moeller, J. and Langway, J. J. Jr. (1969). One thousand centuries of climatic record from Camp Century on the Greenland ice sheet. *Science* 166: 377–381.

Dansgaard, W. (2004). Frozen annals; Greenland Ice Cap Research. Narayana Press.

Dawson, T. E. and Ehleringer, J. R. (1991). Streamside trees that do not use stream water. *Nature* 350: 335–337.

Dawson, T. E., Pausch, R. C. and Parker, H. M. (1998). The role of hydrogen and oxygen stable isotopes in understanding water movement along the soil-plant-atmospheric continuum. *Stable Isotopes: Integration of Biological, Ecological and Geochemical Processes* (H. Griffiths, ed.). Bios Scientific Publishers; Environmental Plant Biology Series. pp. 169–183.

Dawson, T. E. (1998). Fog in the California redwood forest: Ecosystem inputs and use by plants. *Oecologia* 117: 476–485.

Defant, A. (1961) Physical Oceanography, Vol. 1, Pergamon Press.

DeNiro, M. J. and Epstein, S. (1979). Relationship between oxygen isotope ratios of terrestrial plant cellulose, carbon dioxide and water. *Science* 204: 51–53.

Dincer, T. (1968). The use of Oxygen-18 and Deuterium concentrations in the water balance of lakes. *Water Resources Res.* 4: 1289–1306.

Dincer, T., Al-Mughrin, A. and Zimmermann, U. (1974) Study of the infiltration and recharge through the sand dunes in arid zones with special reference to the stable isotopes and thermonuclear tritium. *J. Hydrology* 23: 79–109.

Dincer, T., Hutton, L. G. and Kupee, B. B. J. (1979). Study, using stable isotopes, of surface-groundwater relations and evapo-transpiration in the Okavonga Swamp, Botswana. *Isotope Hydrology 1978*, IAEA, Vienna, pp. 3–26.

Duplessy, J-Cl. (1979). Isotope studies. *Climatic Change* (J. Gribbin, ed.) Cambridge Univ.Press. pp. 46–67.

Edmunds, W. M. (2007). Groundwater as an archive of climatic and environmental change. Chapter 21, *Isotopes in the Water Cycle* (Aggarwal, Froehlich and Gat, eds.), Springer, pp. 341–352.

Ehleringer, J. R. and Dawson, T. E. (1992). Water uptake by plants: perspectives from stable isotope composition. *Plant Cell Environment* 15: 1073–1082.

Ehhalt, D., Roether, W. and Vogel, J. C. (1963). A survey of natural isotopes of water in South Africa. *Radioisotopes in Hydrology*, IAEA, Vienna, pp. 407–415.

Ehhalt, D. (1969). On the deuterium-salinity relationship in the Baltic Sea. *Tellus* 21: 429–435.

Ehhalt, D. (1974). Vertical profiles of HTO, HDO and H_2O in the troposphere. NCAR technical note, NCAR-TN/STR-100.

Eichler, R. (1964). Ueber den Isotopengehalt des Wasserstoffs in Niederschlagsboden und Grundwasser. *Thesis Univ. Heidelberg.*

Eichler, R. (1965). Deuterium Isotopen Geochemie des Grund- und Oberflaechen-wassers. *Geol. Rundschau* 55: 144–155.

Epstein, S., Buchsbaum, R., Lowenstam, H. A. and Urey, H. C. (1951). Carbonate-Water Isotope Scale. *GSA. Bull.* 62: 417–425.

Epstein, S. and Mayeda, T. (1953). Variations of the ^{18}O content of waters from natural sources. *Geochim. Cosmochim. Acta* 4: 213–224.

Epstein, S., Yapp, C. I. and Hall, C. H. (1976). The determination of D/H ratios of nonexchangeable hydrogens in cellulose extracted from aquatic and land plants. *Earth Planet. Science Lett.* 30: 241–251.

Epstein, S., Thompson, P. and Yapp, C. I. (1977). Oxygen and Hydrogen isotopic ratios in plant cellulose. *Science* 198: 1209–1215.

Farquhar, G. D., Barbour, M. M. and Henry, B. K. (1998). Interpretation of oxygen isotope composition of leaf material. *Stable Isotopes: Integration of Biological, Ecological and Geochemical Processes* (H. Griffiths, ed.) Bios Scientific Publishers; Environmental Plant Biology Series. pp. 27–62.

Ferronsky, V. I. and Brezgunov, V. S. (1989). Stable isotopes and Ocean Dynamics. Chapter 1 in *Handbook of Environmental Isotope Geochemistry* (P. Fritz and J. Ch. Fontes, eds.), 3: 1–28.

Fleischer, E., Goldberg, M., Gat, J. R. and Magaritz, M. (1977). Isotope composition of formation waters from deep drillings in southern Israel. *Geochim. Cosmochim. Acta* 41: 511–525.

Foerstel, H. (1978). The enrichment of ^{18}O in leaf water under natural conditions. *Radiat. Environmental Biophys.* 15: 323–344.

Fontes, J. Ch. (1965). Fractionnement isotopique dans l'eau de crystallization du Sulfat du Calcium. *Geol. Rrundschau* 55 :172–178.

Fontes, J. Ch. and Gonfiantini, R. (1967). Fractionnement isotopique de l'hydrogene dans l'eau de crystallisation du gypse. *Comptes Rendues, Academie des Sciences de Paris*, Series D. 265: 4–6.

Fontes, J. Ch. and Gonfiantini, R. (1967). Comportement isotopique au course de l'evaporation de deux basins Sahariens. *Earth Planet. Sci. Letters* 3: 258–266.

Fontes, J. Ch. and Garniers, J. M. (1979). Determination of the initial C-14 activity of the total dissolved carbon. *Water Resources Res.* 15: 399–413.

Fontes, J. Ch. (1981). Palaeowaters. Chapter 12, *Stable Isotope Hydrology: Deuterium and Oxygen-18 in the Water Cycle* (J. R. Gat and R. Gonfiantini, eds.), Technical Report Series No. 210, IAEA, Vienna, pp. 273–302.

Fontes, J. Ch., Yousfi, M. and Allison, G. B. (1986). Estimation of long-term diffusive groundwater discharge in the northern Sahara using stable isotope profiles in soil water. *J. Hydrology* 86: 315–327.

Friedman, I. (1953). Deuterium content of natural waters and other substances. *Geochim. Cosmochim. Acta* 4: 89–103.

Friedman, I., Norton, D. R., Carter, D. B. and Redfield, A. C. (1956). The Deuterium balance of Lake Maracaibo. *Limnology and Oceanography* 1: 230–246.

Friedman I., Machta, L. and Soller, R. (1962). Water vapour exchange between a water droplet and its environment. *J. Geophys. Res.* 67: 2761–2766.

Friedman, I., Redfield, A., Schoen, B. and Harris, J. (1964). The variations of the deuterium content of natural waters in the hydrologic cycle. *Rev. Geophys.* 2: 177–224.

Friedman, I., Benson, C. and Gleason, J. (1991). Isotopic changes during snow metamorphism. *Stable Isotope Geochemistry: A Tribute to Samuel Epstein* (H. P. Taylor, J. R. O'Neill and I. R. Kaplan, eds.), pp. 211–221.

Fritz, P., Cherry, J. A., Weyer, K. V. and Sklash, M. V. (1976). Storm runoff analysis using environmental isotopes and major ions. *Interpretation of Environmental Isotopes and Hydrochemical Data in Groundwater Hydrology* (Proc. Advisory Group Meeting) IAEA Vienna, pp. 111–130.

Fritz, P. and Frape, K. (1982). Comments on the ^{18}O, ^{2}H and chemical composition of saline groundwaters in the Canadian Shield. *Isotope studies of Hydrologic Processes* (E. C. Perry and C. W. Montgommery, eds.), Northern Illinois University Press, pp. 57–63.

Froehlich, K., Grabczak, J. and Rozanski, K. (1988). Deuterium and Oxygen-18 in the Baltic Sea. *Chemical Geology (Isotope Geoscience Section)* 72: 77–83.

Froehlich, K. (2000). Evaluating the water balance of inland seas using isotopic tracers: the Caspian Sea experience. *Hydrological Processes* 14: 1371–1383.

Froehlich, K. F. O, Gonfiantini, R. and Rozanski, K. (2005). Isotopes in lake studies: a historic perspective. *Isotopes in the Water Cycle: Past, Present and Future of a Developing Science* (Aggarwal, Froehlich and Gat, eds.), Springer, pp. 139–150.

Gasse, F. (2005). Isotopic Palaeolimnology. Chapter 22, *Isotopes in the Water Cycle Past, Present and Future of a Developing Science* (Aggarwal, Froehlich and Gat, eds.), Springer, pp. 353–358.

Gat, J. R. and Craig, H. (1965). Characteristics of the air-sea interface determined from isotope transport studies. *Trans. Amer. Geophys. U.* 47: 115.

Gat, J .R. and Tzur, Y. (1967). Modification of the isotopic composition of rainwater by processes which occur before groundwater recharge. *Isotopes in Hydrology,* IAEA, Vienna, pp. 49–60.

Gat, J. R. (1970). Environmental isotope balance of Lake Tiberias. *Isotope Hydrology 1970*, IAEA, Vienna, pp. 109–127.

Gat, J. R. (1971). Comments on the stable isotope method in regional groundwater investigations. *Water Resources Research* 7: 980–993.

Gat, J. R. and Dansgaard, W. (1972). Stable isotope survey of the freshwater occurences in Israel and the Jordan rift valley. *J. Hydrol.* 16: 177–211.

Gat, J. R. and Issar, A. (1974). Desert Isotope Hydrology: water sources of the Sinai Desert. *Geochim. Cosmochim. Acta* 38: 1117–1131.

Gat, J. R. (1975). Elucidating salinization mechanisms by stable isotope tracing of water sources. *Brackish Water as a Factor in Development* (A. Issar, ed.). BG University, pp. 15–24.

Gat, J. R. and Y. Levy (1978). Isotope hydrology of the inland sabkhas in the Bardawil area, Sinai. *Limnol. Oceanogr.* 23: 841–850.

Gat, J. R. (1980). The relationship between surface and subsurface waters: Water quality aspects in areas of low precipitation. *Hydrological Sciences Bull.* 25: 257–267.

Gat, J. R. (1981). Paleoclimate conditions in the Levant as revealed by the isotopic composition of Paleowaters. *Israel Meteorological Research Papers* 3: 13–28.

Gat, J. R. (1983). Precipitation, groundwater and surface waters: Control of climate parameters on their isotope composition and their utilization as palaeoclimatic tools. *Palaeoclimates and Palaeowaters: Application of Environmental Isotope Studies*, IAEA Vienna, pp. 3–12.

Gat, J. R. and Carmi, I. (1987). Effect of climate change on the precipitation patterns and isotopic composition of water in a climate transition zone: Case of the Eastern Mediterranean Sea are. *Influence of Climatic Change on the Hydrological Regime and Water Resources* (Vancouver Symp.) IAHS Publication No. 168: 513–523.

Gat, J. R. and Bowser, C. (1991). The heavy isotope enrichment of water in coupled evaporative systems. *Stable Isotope Geochemistry: A Tribute to Samuel Epstein* (H. P. Taylor, J. R. O'Neill and I. R. Kaplan, eds.) pp. 159–168.

Gat, J. R. and Matsui, E. (1991). Atmospheric water balance in the Amazon basin: an isotopic evapo-transpiration model. *J. Geophys. Res.* 96: 13179–13188.

Gat, J. R., Bowser, C. and Kendall, C. (1994). The contribution of evaporation from the Great Lakes to the continental atmosphere: Estimate based on stable isotope data. *Geophys. Res. Lett.* 21: 557–560.

Gat, J. R. (1995). Stable isotopes of fresh and saline lakes. Chapter 5, *Physics and Chemistry of Lakes* (A. Lerman, D. Imboden and J. Gat, eds.) Springer, pp. 139–165.

Gat, J. R. and Lister, G. S. (1995). The catchment effect on the isotopic composition of lake waters: Its importance in paleo-limnological interpretations. *Problems of Stable Isotopes in Tree Rings, Lake sediments and Peat Bogs as Climatic Evidence* (B. Frensel, ed.). EPC publication. Gustav Fischer V., pp. 1–15.

Gat, J. R. (1996). Oxygen and Hydrogen Isotopes in the Hydrologic Cycle. *Ann. Rev. Earth Planet, Sc.* 24: 225–262.

Gat, J. R., Shemesh, A., Tziperman, E., Hecht A., Georgopoulis, D. and Basturk, O. (1996). The stable isotope composition of waters of the eastern Mediterranean Sea. *J. Geophys. Res.* 101, No. C3: 6441–645l.

Gat, J. R. (1997). The modification of the isotopic composition of meteoric waters at the land/bioshere/atmosphere interface. *Isotope Techniques in the Study of Past and Current Environmental Changes in the Hydrosphere and Atmosphere*, IAEA, Vienna, pp. 153–164.

Gat, J. R. and Rietti-Shati, M. (1999). The meteorological versus the hydrological altitude effect on the isotopic composition meteoric waters. *Isotope Techniques in Water Resources Development and Management* IAEA-Symposium, paper No. SM-361/2.

Gat, J. R. (2000). Atmospheric water balance — The isotopic perspective. *Hydrologic Processes* 14: 1357–1369.

Gat, J. R., Adar, E. and Albert, P. (2001). Inter and intra-storm variability of the isotope composition of precipitation in southern Israel: Are local or large scale factors responsible? *International Conference on the Study of Environmental Change Using Isotope Techniques.* IAEA, Vienna, pp. 41–53.

Gat, J. R. and DeBievre, P. (2002) Making an honest measurement scale out of the oxygen isotope δ-values. *Rapid Comm Mass Spectrometry* 16: 2205–2207.

Gat, J. R., Klein, B., Kushnir, Y., Roether, W., Wernli, H., Yam, R. and Shemesh, A. (2003). Isotope composition of air moisture over the Mediterranean Sea: an index of the air-sea interaction pattern. *Tellus* 55B: 953–965.

Gat, J. R. (2004). The use of environmental isotopes in catchment studies. *Research Basins and Hydrological Planning* (Ru-Ze Xi, Wei-Zu Gu and K. P. Seiler, eds.) Balkema Publ., pp. 3–10.

Gat, J. R. and Airey, P. L. (2006). Stable water isotopes in the atmosphere/ biosphere/lithosphere interface: Scaling up from the local to continental scale under humid and dry conditions. *Global and Planetary Change* 51: 25–33.

Gat, J. R., Yakir, D., Goodfriend, G., Fritz, P., Trimborn, P., Lipp, J., Gev, I., Adar, E. and Waisel, Y. (2007). Stable Isotope Composition of water in desert plants. *Plant and Soil* 298: 31–45.

Gerardo-Abaya, J., D'Amore, F. and Arnorsson, St. (2000). Isotopes for Geothermal Investigations. *Isotopic and Chemical Techniques in Geothermal Exploration, Development and Use* (St. Arnorsson ed.) IAEA, Vienna, pp. 49–65.

Geyh, M. A., Weizu Gu, Jaeckel, D. (1996). Groundwater recharge study in the Gobi Desert, China. *Geo-Wissenschaften* 14: 279–280.

Giauque, W. F. and Johnston, H. L. (1929). An isotope of oxygen: Mass 18. *J. Am. Chem. Soc.* 51:1436–1441.

Gibson, J. J., Edwards, T. W. D. and Prowse, T. D. (1999). Pan-derived isotopic composition of atmospheric water vapour and its variability in northern Canada. *J. Hydrology* 217: 55–74.

Giggenbach, W. F. (1992). Isotopic shifts in waters from geothermal and volcanic systems along convergent plate boundaries and their origin. *Earth Planet. Science Letters* 113: 495–510.

Gonfiantini, R. and Fontes, J. Ch. (1963). Oxygen isotope fractionation in the water of crystallization of gypsum. *Nature* 200: 644–646.

Gonfiantini, R. (1965). Effeti isotopici nell evaporazione di acque salate (Isotope effects in salt water evaporation). *Attidella Societa Toscana di Scienza Naturali, serie A*72: 550–569.

Gonfiantini, R., Gratziu, S. and Tongiorgi, E. (1965). Oxygen isotope composition of water in leaves. *Isotopes and Radiation in Soil Plant Nutrition Studies.* Tech Rep. Series No. 206. IAEA, Vienna, pp. 405–410.

Goodfriend, G. A., Magaritz, M. and Gat, J. R. (1989). Stable isotope composition of land snail body water and its relation to environmental waters and shell carbonate. *Geochim. Cosmochim. Acta* 53: 3215–3221.

Gourcy, L. L., Groening, M. and Aggarwal, P. K. (2005). Stable Oxygen and Hydrogen isotopes. Chapter 4, *Isotopes in the Water Cycle: Past, Present and Future of a Developing Science.* (Aggarwal, Froehlich and Gat, eds.), Springer: pp. 39–51.

Gvirtzman, H. and Magaritz, M. (1986). Investigation of water movement in the unsaturated zone under an irrigated area using environmental tritium. *Water Resources Research* 22: 635–642.

Gvirtzman, H. and Magaritz, M. (1990). Water and anion transport of the unsaturated zone traced by environmental tritium. *Inorganic Contaminants in the Vadose Zone* (B. Bar-Yosef, N. J. Barrow and J. Goldschmidt, eds.), Springer. Ecological studies 74: 190–198.

Hagemann, R., Nief, G. and Roth, E. (1970). Absolute isotope scale for deuterium analysis of natural waters; Absolute D/H ratio for SMOW. *Tellus* 22: 712–715.

Halevy, E. (1970). The relationship between isotopic composition of precipitation and lysimeter percolates. Report to the IAEA, pp. 10.

Harmon, R. S. (1979). An isotopic study of groundwater seepage in the central Kentucky karst.*Water Resources Research* 15: 476–480.

Henderson-Sellers, A. (2006). Improving land-surface parametrization schemes using stable water isotopes: Introducing the iPILPS initiative. *Global and Planetary Change* 51: 3–24.

Hitchon, B. and Friedman, I. (1969). Geochemistry and origin of formation waters in the western Canada sedimentary basin: I. Stable isotopes of hydrogen and oxygen. *Geochim. Cosmochim. Acta* 33: 1321–1349.

Hoffmann, G., Werner, M. and Heimann, M. (1998). The water isotope module of the ECHAM atmospheric general circulation model — A study on time scales from days to several years. *J.Geophys. Res.* 103: 16871–16896.

Hoffmann, G., Cuntz, M., Jouzel, J. and Werner, M. (2005). How much climatic information do water isotopes contain. Chapter 19, *Isotopes in the Water Cycle: Past, Present and Future of a Developing Science* (P. Aggarwal, J. R. Gat and K. Froehlich, eds.), pp. 303–320.

Horita, J. (1988) Hydrogen isotope analysis of natural waters using an H_2-water equilibration method: A special implication to brines. *Chem. Geol. (Isotope Geoscience Section)* 72: 89–94.

Horita, J. and Gat, J. R. (1988) Procedure for the hydrogen isotope analysis of water from concentrated brines. *Chem. Geol. (Isotope Geoscience Section)* 72: 85–88.

Horita, J. (1989). Analytical aspects of stable isotopes in brines. *Chem. Geol. (Isotope Geoscience Section)* 79: 107–112.

Horita, J. and Gat, J. R. (1989) Deuterium in the Dead Sea: Remeasurement and implications for the isotope activity correction in brines. *Geochim. Cosmochim. Acta* 53: 131–133.

Horita, J., Ueda, A., Mizukami, A. and Takatori, I. (1989). Automatic δD and $\delta^{18}O$ analysis of multi-water samples using H_2 and CO_2-water equilibrium methods with a common equilibration set-up. *Int. J. Rad. Applied Instrum., Part A.* 40: 801–805.

Horita, J., Cole, D. R. and Wesolowski, D. J. (1993). The activity-composition relationship of oxygen and hydrogen isotopes in aqueous salt solutions. II: Vapour-liquid water equilibrium of mixed salt solutions from 50 to 100°C and geochemical implications. *Geochim. Cosmochim. Acta* 57: 4703–4711.

Horita, J., Cole, D. R. and Wesolowski, D. J. (1995). The activity-composition relationship of oxygen and hydrogen isotopes in aqueous salt solutions. III: Vapour-liquid water equilibration of NaCl solutions to 350°C. *Geochim. Cosmochim. Acta* 59: 1139–1151.

Horita, J. (2005). Saline Waters. In: *Isotopes in the Water Cycle: Past, Present and Future of a Developing Science* (Aggarwal, Froehlich and Gat, eds.), Springer: Chapter 17: 271–287.

Horovitz, A. and Gat, J. R. (1984). Floral and isotopic indications for possible summer rains in Israel during wetter periods. *Pollen and Spores* 26: 61–68.

Imboden, D. and Wuest, A. (1995). Mixing Mechanisms in Lakes. In: *Physics and Chemistry of Lakes* (A. Lerman, D. Imboden and J. R. Gat, eds.), Springer: Chapter 4: 81–138.

Ingraham, N. L. and Matthews, R. A. (1988). Fog drip as a source of groundwater recharge in northern Kenya. *Water Resources Research* 24: 1406–1410.

Ingraham, N. L. and Matthews, R. A. (1990). A stable isotope study of fog: The Point Reyes Peninsula, California, USA. *Chemical Geology (Isotope Geoscience Section)* 80: 281–290.

Issar, A., Bein, A. and Michaeli, A. (1972). On the ancient waters of the upper Nubian Sandstone aquifer in central Sinai and southern Israel. *J. Hydrology* 17: 353–374.

Issar, A. S., Gat, J. R., Karnieli, A., Nativ, R. and Mazor, E. (1985). Groundwater formation under desert conditions. *Stable and Radioactive Isotopes in the Study of the Unsaturated Soil Zone,* IAEA-Tecdoc 357, Vienna. pp. 35–54.

Jacob, H. and Sonntag, Ch. (1991). An 8-year record of the seasonal variations of ^{2}H and ^{18}O in atmospheric water vapour and precipitation in Heidelberg, Germany. *Tellus* 43B: 291–300.

Joussaume, J., Sadourny, R. and Jouzel, J. (1984). A general circulation model of water isotope cycles in the atmosphere. *Nature* 311: 24–29.

Jouzel, J., Merlivat, L. and Roth, E. (1975). Isotopic study of hail. *J. Geophys. Res.* 80: 5015–5030.

Jouzel, J. and Merlivat, L. (1984). Deuterium and oxygen-18 in precipitation: Modeling of the isotope effects during snow formation. *J. Geophys. Res.* 89: 11749–11757.

Jouzel, J., Merlivat, L. and Federer, B. (1985). Isotopic study of hail: The δD − δ^{18}O relationship and the growth history of hailstones. *Q. J. Royal Meteor. Soc.* 111: 495–516.

Jouzel, J., Russel, G. I., Suozzo, R. J., Koster, R. D., White, J. W. C. and Broecker. W. S. (1987). Simulations of the HDO and ^{18}O atmospheric cycles using the NASA GISS General Circulation Model. *J. Geophys. Res.* 92: 14739–14760.

Jouzel, J., Froehlich, K. and Schotterer, U. (1997). Deuterium and oxygen-18 in present-day precipitation and modelling. *Hydrological Sciences Journal* 42: 747–763.

Kaplan, I. R., Zhang, D. and Gat, J. R. (1995). The source of water used by streambed vegetation from Owen Valley in the Sierra Nevada. *EOS Suppl.* pp. F214.

Kendall, C. (1993). Impact of isotopic heterogeneity in shallow systems on modeling of stormflow generation. Ph.D thesis, Dpt. of Geology, University of Maryland, pp. 270.

Kendall, C. and McDonnell, J. J. (1993). Effect of intrastorm isotopic hetereogeneities of rainfall, soil water and groundwater on runoff modeling. *Tracers in Hydrology*. IAHS 215: 41–48.

Kendall, C. and Coplen, T. (2001). Distribution of oxygen–18 and deuterium in river waters across the United States. *Hydrol. Processes* 15: 1363–1393.

Kerstel, E. R. Th. and Meijer, H. A. J. (2005). Optical isotope ratio measurements in Hydrology. *Isotopes in the Water Cycle: Past, Present and Future of a Developing Science* (Aggarwal, Froehlich and Gat, eds.), Springer: pp. 109–123.

Kharaka, Y. K. and Carothers, W. W. (1986). Oxygen and Hydrogen isotope geochemistry of deep basin brines. Chapter 8, *Handbook of Environmental Isotope Geochemistry* (P. Fritz and J. Ch. Fontes, eds.), Elsevier 2: 305–360.

Kharaka, Y. K. and Mariner, R. H. (2005). Geothermal Systems. Chapter 16, In: *Isotopes in the Water Cycle: Past, Present and Future of a Developing Science* (Aggarwal, Froehlich and Gat, eds.), Springer: pp. 243–270.

Kirschenbaum, I. (1951). Physical Properties and Analysis of Heavy Water. *National Nuclear Energy Series*, Div. III, 4a, pp. 438.

Kloppmann, W., Girard, J. P. and Negrel, P. (2002). Exotic stable isotope composition of saline waters and brines from crystalline basements. *Chem. Geology* 184: 49–70.

Krabbenhoft, D. P., Bowser, C. J., Anderson, M. P. and Valley, J. W. (1990). Estimating groundwater exchange with lakes, 1. The stable Isotope Mass Balance Method. *Water Resources Res.* 26: 2445–2453.

Krabbenhoft, D. P., Bowser, C. J., Kendall, C. and Gat, J. R. (1994). Use of Oxygen-18 and Deuterium to assess the hydrology of Groundwater-Lake systems. *Environmental Chemistry of Lakes and Reservoirs* (L. A. Baker, ed.) ACS Advances in Chemistry Series No. 237, pp. 67–90.

Landais, A., Barkan, E. and Luz, B. (2008). The record of $\delta^{18}O$ and ^{17}O-excess in ice from Vostok Antarctica during the last 150,000 years. *Geophys. Res. Lett.* 35: L02709, 5p.

Landwehr, J. M. and Coplen, T. B. (2006). Line-conditioned excess: A new method for characterizing stable hydrogen and oxygen isotope ratios in hydrologic systems. *Isotopes in Environmental Studies*. IAEA-CSP-26, Vienna, pp. 132–135.

Langbein, W. B. (1961). Salinity and hydrology of closed lakes. *USGS Professional Papers* # 412.

Lawrence, J. R., Gedzelman, S. D., Zhang, X. and Arnold, R. (1998). Stable isotope ratios of rain and vapor in 1995 hurricanes. *J. Geophys. Res.* 103: 11381–11400.

Lee, K. S., Werner, D. B. and Lee, I. (1999). Using H and O isotopic data for estimating the relative contribution of rainy and dry season precipitation to groundwater: Example for Cheju Island, Korea. *J. Hydrology* 222: 65–74.

Leguy, C., Rindsberger, M., Zangwil, A., Issar, A. and Gat, J. R. (1983). The relation between the Oxygen-18 and Deuterium contents of rainwater in the Negev Desert and air mass trajectories. *Isotope Geosciences* 1: 205–218.

Lerman, A. and Clauer, N. (2007). Stable Isotopes in the Sedimentary Record. *Treatise on Geochemistry*, Elsevier, 7: 1–55.

Levin, M., Gat, J. R. and Issar, A. (1980). Precipitation, flood and groundwaters of the Negev Highlands: An isotopic study of desert hydrology. *Arid-Zone Hydrology: Investigations with Isotope Techniques.* Proc. IAEA Advisory Group Meeting, IAEA-AG-158/1, pp. 3–22.

Liu, K. K. (1984). Hydrogen and Oxygen isotope compositions of meteoric waters from the Tatun Shan area, northern Taiwan. *Academia Sinica* 4: 159–175.

Lloyd, R. M. (1966). Oxygen isotope enrichment of sea water by evaporation. *Geochim. Cosmochem. Acta* 30: 801–814.

Longinelli, A. and Nuti, S. (1965). Oxygen isotope composition of phosphate and carbonate from living and fossil marine organisms. *Stable Isotopes in Oceanographic Studies and Paleotemperature* (E. Tongiorgi, ed.) Lab. di Geologia Nucleare, Pisa. pp. 183–187.

Lord Rayleigh (1902). On the distillation of binary mixtures. *Phil. Magaz. 4* (series 6): pp. 521–537.

Lucas, L. L. and Unterweger, M. P. (2000). Comprehensive review and critical evaluation of the half-life of tritium. *J. Res. Natl. Institute of Standards and Technology* 104 (4): 541–549.

Luz, B., Barkan, E., Bender, M. L., Thiemens, M. and Boerings, K. A. (1999). Triple-isotope composition of atmospheric oxygen as a tracer of biosphere production. *Nature* 400: 547–550.

Magaritz, M. and Heller, J. (1980). A desert migration indicator — Oxygen isotope composition of land snail shells. *Paleogeograph. Paleoclimatol. Paleoecology* 32: 153–162.

Magaritz, M. and Gat, J. R. (1981). Review of the natural abundance of Hydrogen and Oxygen isotopes. Chapter 5, *Stable Isotope Hydrology: Deuterium and Oxygen-18 in the Water Cycle* (J. R. Gat and R. Gonfiantini, eds.), IAEA, Vienna. pp. 85–102.

Majoube, M. (1971). Fractionnement en oxygene-18 et en deuterium entre l'eau et sa vapeur. *J. Chim. Phys.* 68: 1423–1436.

McKinney, C. R., McCrea, J. M., Epstein, S., Allen, H. A. and Urey, H. C. (1950). Improvements in mass spectrometers for the measurements of small differences in isotope abundance ratios. *Rev. Sci. Instrum* 21: 724–730.

McKenzie J. A., Hsu, K. J. and Schneider, J. F. (1980). Movement of sub-surface waters under the Sabkha, Abu Dhabi and its relation to evaporative dolomite genesis. *Concepts and Models of Dolomitisation*. Soc. Econ. Paleoton. Miner. Special Publ. 28, pp. 11–30.

Macklin, W. C., Merlivat, L. and Stevenson, C. M. (1970). The analysis of a hailstone. *Quart. J. Royal Meteorol. Soc.* 96: 472–486.

Manabe, S. (1969). Climate and ocean circulation: 1, the atmospheric circulation and the hydrology of the Earth's surface. *Monthly Weather Review* 97: 739–805.

Martinelli, L. A., Victoria, R. L., Matsui, E., Richey, J. E., Forsberg, B. R. and Mortatti, J. (1989). The use of Oxygen isotopic composition to study water dynamics in Amazon floodplain lakes. *Isotope Hydrology Investigations in Latin America.* IAEA Tecdoc 502, pp. 91–101.

Martinelli, L. A., Gat, J. R., DeCamargo, P. B., Lara, L. L. and Ometto, P. H. B. (2004). The Piracicaba River Basin: Isotope Hydrology of a tropical river basin under anthropogenic stress. *Isotopes in Environmental and Health Studies* 40: 45–56.

Matsui, E., Salati, E., Ribeiro, M., Reis, C. M., Tancredi, A. and Gat, J. R. (1983). Precipitation in the central Amazon basin: The isotopic composition of rain and atmospheric moisture at Belem and Manaus. *Acta Amazonica* 13: 307–369.

Mazor, E. (1972). Palaeotemperatures and other hydrological parameters deduced from noble gases dissolved in groundwaters, Jordan Rift Valley, Israel. *Geochim. Cosmochim. Acta* 36: 1321–1336.

Meijer, H. A. and Li, W. J. (1998). The use of electrolysis for accurate $\delta^{17}O$ and $\delta^{18}O$ isotope measurements in water. *Isotopes in Environmental and Health Studies* 34: 349–369.

Merlivat, L. and Nief, G. (1967). Fractionnement isotopique lors de changements d'etat solide-vapeur et liquide-vapeur de l'eau a des temperatures inferieures a 0°C. *Tellus* 19: 122–129.

Merlivat, L. and Coantic, M. (1975). Study of mass transfer at the air-water interface by an isotopic method. *J. Geophys. Res.* 80: 3455–3464.

Merlivat, L. (1978). Molecular diffusivities of $H_2^{16}O$, $HD^{16}O$ and $H_2^{18}O$ in gases. *J. Chem. Phys.* 69: 2864–2871.

Merlivat, L. and Jouzel, J. (1979). Global climatic interpretation of the deuterium-oxygen-18 relationship for precipitation. *J. Geophys. Res.* 84: 5029–5033.

Molion, L. C. B. (1987). Micro-meteorology of an Amazonian rain-forest. *The Geo-Physiology of Amazonia*; vegetation and climate interactions (R. E. Dickinson, ed.) John Wiley for the U.N. University, New-York, Chapter 14: pp. 255–270.

Moser, H. and Stichler, W. (1971). Die Verwendung des Deuterium und Sauerstoff-18 Gehalts bei Hydrologischen Untersuchungen. *Geolog. Bavarica* 64: 7–35.

Moser, H. and Stichler, W. (1980) Environmental Isotopes in Ice and Snow. Chapter 4, *Handbook of Environmental Isotope Geochemistry* (P. Fritz and J. Ch. Fontes, eds.), 1: 141–178.

Muennich, K. O. and Vogel, J. C. (1962). Untersuchungen an pluvialen Wassern der Ost-Sahara. *Geol. Rundschau* 52: 611–624.

Nier, A. O. (1947). A mass spectrometer for isotope and gas analysis. *Rev. Scient. Instruments* 18: 398–404.

Ninari, N. and Berliner, P. R. (2002). The role of dew in the water and heat balance of bare Loess soil in the Negev Desert; Quantifying the actual dew deposition on the soil surface. *Atmos. Res.* 64: 325–336.

Nissenbaum, A., Lifshitz, A. and Stepek, A. (1974). Detection of citrus fruit adulteration using the distribution of natural isotopes. *Lebensm. Wiss. Technol.* 7: 152–154.

O'Neil, J. R. (1968). Hydrogen and Oxygen isotope fractionation between ice and water. *J. Phys. Chem.* 72: 3683–3684.

O'Neil, J. R. and Truesdell, A. H. (1991). Oxygen isotope fractionation studies of solute-water interactions. *Stable Isotope Geochemistry: A Tribute to Samuel Epstein* (H. P. Taylor, J. R. O'Neill and I. R. Kaplan, eds.), pp. 17–25.

Oestlund, H. G. and Berry, X. (1970). Modification of atmospheric tritium and water vapor by Lake Tahoe. *Tellus* 22: 463–465.

Ohte, N., Koba, K., Yoshikawa, K., Sugimoto, A., Matsuo, N., Kabeya, N. and Wang, L. (2003). Water utilization of natural and planted trees in the semiarid desert of Inner Mongolia, China. *Ecological Applications* 13: 337–351.

Pionke, H. B. and DeWalle, D. R. (1992). Intra- and inter-storm [18]O trends for selected rainstorms in Pennsylvania. *J. Hydrology* 138: 131–143.

Ramesh, R. and Sarin, M. M. (1992). Stable isotope survey of the Ganga (Ganges) river system. *J. Hydrology* 139: 49–62.

Redfield, A. C. and Friedman, I. (1965). Factors affecting the distribution of deuterium in the ocean. *Symposium on Marine Geochemistry.* Univ. Rhode Island, Occasional Publications 3: 149–168.

Redfield, A. C. and Friedman, I. (1969). The effect of meteoric water, melt water and brine on the composition of polar sea water. *Deep Sea Res.*, Suppl. to Vol 16, pp. 197–214.

Rindsberger, M. and Magaritz, M. (1983). The relation between air-mass trajectories and the water isotope composition of rain in the Mediterranean Sea region. *Geophys. Res. Letters* 10: 43–46.

Rindsberger, M., Jaffe, Sh., Rahamim, Sh. and Gat, J. R. (1990). Patterns of the isotopic composition of precipitation in time and space: Data from the Israeli storm water collection program. *Tellus* 42B: 263–271.

Rozanski, K., Araguas-Araguas, L. and Gonfiantini, R. (1993). Isotopic patterns in modern global precipitation. *Climate Change in the Continental Isotopic Record* (P. K. Swart, K. L. Lohman, J. A. Mckenzie and S. Savin, eds.) Geophys. Monograph 78: 1–37.

Rozanski, K., Froehlich, K. and Mook, W. G. (2001). Lakes and Reservoirs. Environmental Isotopes in the hydrological cycle; Vol. III: Surface Waters, IHP-V, Technical Documents in Hydrology, No 39, Vol III, UNESCO, pp. 59–92.

Rozanski, K. (2005). Isotopes in Atmospheric Moisture. *Isotopes in the Water Cycle: Past, Present and Future of a Developing Science* (Aggarwal, Froehlich and Gat, eds.), Springer: Chapter 18, pp. 291–302.

Salati, E., Dal'Ollio, A., Matsui, E. and Gat, J. R. (1979). Recycling of water in the Amazon Basin: An isotopic study. *Water Resources Res.* 15: 1250–1258.

Sauzay, G. (1974). Sampling of lysimeters for environmental isotopes of water. *Isotope Techniques in Groundwater Hydrology, 1974.* IAEA, Vienna, pp. 61–68.

Savin, S. M. and Epstein, S. (1970). The oxygen and hydrogen isotope geochemistry of clay minerals. *Geochim. Cosmochim. Acta* 34: 25–42.

Schmidt, G. A., Bigg, G. R. and Rohling, E. J. (1999). Global seawater oxygen-18 database. URL://www.giss.nasa.gov/data/o18data/

Scholl, M. A., Gingerich, S. B. and Tribble, G. W. (2002). The influence of micro-climates and fog on stable isotope signatures used in interpretation of regional hydrology: East Maui, Hawaii. *J. Hydrology* 264: 170–184.

Scholl, M. A., Eugster, W. and Burkard, R. (In press, 2009). Understanding the role of fog in forest hydrology: Stable isotopes as tools for determining input and partitioning of cloud water in Montane Forests. *Mountains in the Mist: Science for Conserving and Managing Tropical Montane Cloud Forests* (L. A. Bruijnzeel *et al.*, eds.). Cambridge University Press.

Schwarcz, H. P. (1986). Geochronology and Isotope Geochemistry of speleothems. Chapter 7, *Handbook of Environmental Isotope Geochemistry* (P. Fritz and J. Ch. Fontes, eds), 2: 271–303.

Shank, W. C., Bohlke, J. K. and Seal, R. R. (1995). Stable isotopes in mid-ocean ridge hydrothermal systems: Interactions between fluids, minerals and organisms. *Seafloor Hydrothermal Systems: Physical, Chemical, Biological and Geological Interactions.* AGU Geophysical Monograph 91: 194–221.

Simpson, E. S., Thornd, D. B. and Friedman, I. (1972). Distinguishing seasonal recharge to groundwater by deuterium analysis in Southern Arizona. *World Water Balance*, UNESCO-WMO, pp. 623–633.

Simpson, B. and Carmi, I. (1983). The hydrology of the Jordan tributaries (Israel): Hydrographicand isotopic investigation. *J. Hydrology* 62: 225–249.

Sklash, M. G., Farvolden, R. N. and Fritz, P. (1979). A conceptional model of watershed response to rainfall, developed through the use of Oxygen-18 as a natural tracer. *Can. J. Earth Sc.* 13: 2731–283.

Sofer, Z. and Gat, J. R. (1972). Activities and concentrations of Oxygen-18 in concentrated aqueous salt solutions: Analytical and geophysical implications. *Earth Planet. Science Lett.* 15: 232–238.

Sofer, Z. and Gat, J. R. (1975). The isotopic composition of evaporating brines: Effects of the isotope activity ratio in saline solutions. *Earth Planet. Science Lett.* 26: 179–186.

Sofer, Z. (1978). Isotopic composition of hydration water in gypsum. *Geochim. Cosmochim. Acta* 42: 1141–1149.

Sonntag, C., Klitsch, E., Lohnert, E. P., Muennich, K. O., Junghans, C., Thorweihe, U., Weistroffer, K. and Swailem, F. M. (1978). Paleoclimate information from D and [18]O in ^{14}C-dated North Saharian groundwaters; groundwater formation from the past. *Isotope Hydrology*, IAEA, Vienna, pp. 569–580.

Sonntag, C., Christmann, D. and Muennich, K. O. (1985). Laboratory and field experiments on infiltration and evaporation of soilwater by means of deuterium and Oxygen-18. *Stable and Radioactive Isotopes in the Study of the Unsaturated Soil Zone*, IAEA-Tecdoc 357, Vienna. pp. 145–160.

Steckel, F. and Szapiro, S. (1963) Physical Properties of Heavy-Oxygen Water; Part 1: Density and thermal expansion. *Trans. Faraday Soc.* 59: 331.

Stiller, M., Kaushansky, P. and Carmi, I. (1983). Recent climate changes recorded by the salinity of pore waters in the Dead Sea sediments. *Hydrobiologia* 103: 75–79.

Stuiver, M. (1970). Oxygen and Carbon isotope ratios of freshwater carbonates as climatic indicators. *J. Geophys. Res.* 75: 5247–5257.

Sugimoto, A., Higuchi, K. and Kusakabe, M. (1988). Relationship between δD and $\delta^{18}O$ values of falling snow particles from a separate cloud. *Tellus* 40B: 205–213.

Sugimoto, A. and Higuchi, K. (1989). Oxygen isotopic variation of falling snow particles with time during the lifetime of a convective cloud: Observation and modeling. *Tellus* 41B: 511–523.

Sugimoto, A., Naito, D., Yanagisawa, N., Ichinayagi, K., Kurita, N. Kubota, J., Kotake, T., Ohata, T., Maximov, T. C. and Fedorov, A. N. (2003). Characteristics of soil moisture in permafrost observed in East Siberian taiga with stable isotopes of water. *Hydrologic Processes* 17: 1073–1092.

Sverdrup, H. U. (1951). Evaporation from the oceans. *Compendium of Meteorology. American Meteorological Soc.*, pp. 1071–1081.

Swart, P. K. (1991). The hydrogen and oxygen isotope composition of the Black Sea. *Deep Sea Research, Part A* 38 (Suppl 2), pp. 761–772.

Switsur, R. and Waterhouse, J. (1998). Stable isotopes in tree ring cellulose. Chapter 18, *Stable Isotopes: Integration of Biological, Ecological and Geochemical Processes* (H. Griffiths,). Bios Scientific Publishers; Environmental Plant Biology Series. pp. 303–321.

Szapiro, S. and Steckel, F. (1967). Physical Properties of Heavy-Oxygen Water; Part 2: Vapour Pressure. *Trans. Faraday Soc.* 63: 883–894.

Taube, H. (1954). Use of oxygen isotope effects in the study of the hydration of ions. *J. Phys. Chem.* 58: 523–528.

Taylor, C. B. (1972). The vertical variations of isotopic compositions of tropospheric water vapour over continental Europe and their relation to tropospheric structure. *Report INS-R-107*, Lower Hutt, New Zealand.

Taylor, S., Feng, X., Kirchner, J. W., Osterhuber, R., Klaue, B. and Renshaw, C. E. (2001). Isotopic evolution of a seasonal snowpack and its melt. *Water Resources Res.* 37: 759–769.

Thiemens, M. H., Jackson, T., Zipf, E. C., Erdman, P. W. and van Egmond, C. (1995). Carbon Dioxide and Oxygen isotope anomalies in the Mesosphere and Stratosphere. *Science* 270: 969–971.

Thompson, L. G. and Davis, M. E. (2007). Stable isotopes through the Holocene as recorded in low-latitude ice cores. Chapter 20, *Isotopes in the Water Cycle: Past, Present and Future of a Developing Science* (Aggarwal, Froehlich and Gat, eds.), Springer, pp. 321–339.

Truesdell, A. H., Nathenson, M. and Rye, R. O. (1977). The effects of subsurface boiling and dilution on the isotopic compositions of Yellowstone Thermal Waters. *J. Geophys. Res.* 82: 3694–3704.

Umnikrishna, P. V., McDonnell, J. J. and Kendall, C. (2002). Isotope variations in a Sierra Nevada snowpack and their relation to meltwater. *J. Hydrology* 260: 38–57.

Urey, H. C., Brickwedde, F. G. and Murphy, G. M. (1932). A hydrogen isotope of mass 2. *Phys. Reviews* 39: 1–15.

Vannote, R. I. *et al.* (1980). The River Continuum Concept. *Can. J. Fish. Aquatic Sciences* 37: 130–137.

Vogel, J. and VanUrk, H. (1975). Isotope composition of groundwaters in semi-arid regions of South Africa. *J. Hydrol.* 25: 23–36.

Wang, J. H., Robinson, C. and Edelman, I. S. (1953). Self diffusion and structure of liquid water: III: Measurement of the self diffusion of liquid water with ^{2}H, ^{3}H and ^{18}O as tracers. *J. Amer. Chem. Soc.* 75: 466–470.

Wang, X. F. and Yakir, D. (1995). Temporal and spatial variations in the O-18 content of leaf water in different plant-species. *Plant Cell Environ.* 18: 1377–1385.

Wernli, H. and Davies, H. C. (1997). A Lagrangian-based analysis of extratropical cyclones. I: The method and some applications. *Q. J. Roy. Meteorolog. Soc.* 123: 467–489.

Wershaw, R. L., Friedman, I., Heller, S. J. and Frank, P. A. (1970). Hydrogen isotopic fractionation of water passing through trees. *Advances of Organic Geochemistry* (G. D. Hobson, ed.). Pergamon Press, Oxford, pp. 55–67.

Worden, J., Noone, D., Bowman, K. *et al.* (2007). Importance of rain evaporation and continential convection in the tropical water cycle. *Nature* 445: 528–532.

Yair, A. (1990). Runoff generation in a sandy area — The Nizzana sands, western Negev, Israel. *Earth Surface Process. Landforms* 15: 597–609.

Yakir, D., DeNiro, M. J. and Gat, J. R. (1990). Natural deuterium and Oxygen-18 enrichment in leaf water of cotton plants grown under wet and dry conditions: Evidence for water compartmentation and its dynamics. *Plant, Cell and Environment* 13: 49–56.

Yakir, D. and Yechieli, Y. (1995). Plant invasion of newly exposed hyper-saline Dead Sea shores. *Nature* 374: 803–805.

Yakir, D. (1998). Oxygen-18 of leaf water: A crossroad for plant-associated isotopic signals. Ch. 10, *Stable Isotopes: Integration of Biological, Ecological and Geochemical Processes* (H. Griffiths, ed.). Bios Scientific Publishers; Environmental Plant Biology Series. pp. 147–168.

Yapp, C. J. (1979). Oxygen and Carbon isotope measurements of land snail shell carbonate. *Geochim. Cosmochim. Acta* 43: 629–635.

Yurtsever, Y. (1975). Worldwide survey of stable isotopes in precipitation. *Rep. Isotope Hydrology Section, IAEA*, Vienna. 44 pp.

Yurtsever, Y. and Gat, J. R. (1981). Atmospheric Waters. Chapter 6, *Stable Isotope Hydrology: Deuterium and Oxygen-18 in the Water Cycle* (Gat and Gonfiantini, eds.) IAEA Technical Report Series No. 210, pp. 103–142.

Zieborak, K. (1966). Boiling temperatures and vapour pressures of $H_2O–D_2O$ mixtures and their azeotropes. *Z. Physik. Chemie* 231: 248–258.

Zimmermann, U., Ehhalt, D. and Muennich, K.O. (1967). Soil water movement and evapo–transpiration: Changes in the isotopic composition of water. *Isotopes in Hydrology*, IAEA, Vienna, pp. 567–584.

Zimmermann, U., Munnich, K. O. and Roether, W. (1967) Downward movement
    of soil moisture traced by means of hydrogen isotopes. *Isotope Techniques in
    the Hydrologic Cycle*, Amer. Geoph. Union, Geophys. Monographs 11: 28–36.
Zuber. A. (1983). On the environmental isotope method for determining the water
    balance components of some lakes. *J. Hydrology* 61: 409–427.

Appendix

Tritium in the Water Cycle

The hydrogen isotope of mass 3 (^{3}H, also termed Tritium, T) is radioactive with a half-life of 12.32 years (Lucas and Unterweger, 2000). From measurements on natural waters in the 1940's, it was inferred that Tritium is found naturally at an abundance of T/H in the order of magnitude of $1/10^{18}$(Grosse et al., 1951), which led to the introduction of the "Tritium Unit" where 1 TU (i.e. ^{3}H/^{1}H $= 10^{-18}$) is equivalent to 7.2 dpm per liter of water, which is 3.2 pCurie/L or 0.118 Bequerel/L. Tritium decays by β emission, resulting in the isotope ^{3}He as its daughter product:

$$^3\text{H} \rightarrow (\beta) \rightarrow {}^3\text{He} + \text{e}$$

In analogy to ^{14}C, Libby and his coworkers inferred that the source for the tritiated atmospheric waters is the interaction of high energy cosmic radiation with atmospheric constituents according to the scheme:

$$^{14}\text{N} + \text{n} \rightarrow {}^{12}\text{C} + {}^3\text{H}$$

As reviewed by Lal and Peters (1967), most of the Tritium thus formed in the upper atmosphere is then oxidised and incorporated into water in the form of the tritiated water molecule ^{3}H^{1}HO, but Tritium is also found in hydrogen gas and methane in the atmosphere. Other natural sources of Tritium by fission and interactions with thermal neutrons in the lithosphere are considered relatively negligible in comparison.

There are very few measurements of Tritium levels in atmospheric waters before the anthropogenic effects perturbed the natural system (to be described below). Measurements in precipitation and surface waters by Kaufmann and Libby (1954) in Chicago and other northern hemisphere locations indicated values between 4 to 25 TUs, with the lower values at locations close to the ocean.

Since the tritiated water molecule can be expected to be an almost ideal tracer for the movement of water, it was foreseen (Libby, 1953) that the comparison of the actual Tritium value in a water body with that of the precipitation could be a measure of the residence time of meteoric waters in the system, scaled by the decay rate of the Tritium. The possible use for dating agricultural products, wine in particular, by comparing the actual value with the supposedly original Tritium concentration was also realised. Changes of the Tritium concentration due to an isotope effect accompanying the phase transitions of water (as was discussed for the stable isotopes in Chapter 3) is also to be anticipated for the tritiated water molecules. Indeed, as discussed by Bigeleisen (1962), the Tritium isotope effects are expected to be almost double those of the deuterated molecules. However, this change is completely negligible in comparison to the change imposed by the radioactive decay. An obvious shortcoming for widespread application of this methodology was the lengthy measurement procedure which necessitated electrolytic pre-enrichment of the Tritium in the sample before conducting the counting of the decay rate using an anti-coincidence shield for the proportional counter, in view of the very low natural Tritium levels. With the use of the scintillation counter, the procedure has now been simplified.

The continuing monitoring of the Tritium levels in precipitation in Chicago (Buttlar and Libby, 1955) and Ottawa (Brown and Grummitt, 1956) showed a dramatic increase by an order of magnitude as from March 1954. (An earlier short spike was noted already in 1952). Continued worldwide monitoring by the GNIP program recorded episodal peaks of increasing amplitude, reaching values as high as 2500 TU in 1964. The correlation with commensurate peaks of fission products and the timing of thermo-nuclear atmospheric tests in 1954, 1955 and 1958 and again during 1961–62 made it obvious that these peaks were the result of the dissemination of the products of these tests. Since the establishment of the (atmospheric) weapon testing moratorium in 1963, the levels have been decreasing rather rapidly due to the radioactive decay of the atmospheric reservoir and the rainout and exchange with the oceanic systems, so that levels close to pre-bomb times have been approached in recent years.

The data, based on the records of the IAEA-GNIP program, are summarised in Fig. A.1. The most notable feature is a yearly cycle of maximum concentrations in spring and summer and a winter minimum, with typical concentration ratios of 2.5–6 between maximum and minimal values in any year. This annual cycle is superimposed upon the long-term changes which

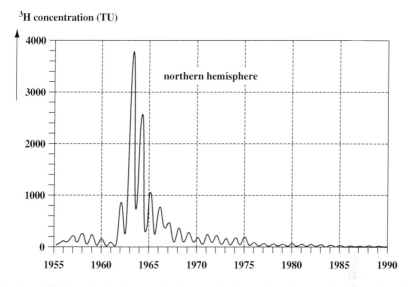

Fig. A.1.   Tritium levels of atmospheric water in the northern hemisphere (based on the GNIP database).

have ranged over 3 orders of magnitudes since 1952. There is a marked latitudinal effect: concentration were highest north of the 30th parallel, with values much lower at tropical and low latitude stations (Athavale *et al.* 1967) during the years following the thermo-nuclear injections into the atmosphere, a consequence of the location of most of these injections in the northern hemisphere and the low inter-hemispheric exchange of the tracer. It is only following the cessation of nuclear atmospheric testing and the subsequent declining Tritium values that the Tritium concentrations in the precipitation of the northern and southern hemispheres are comparable.

The pattern described above obviously is the result primarily of the interaction mode between the stratospheric and tropospheric air masses. Whereas throughout most of the year, these two spheres are relatively isolated by the tropopause layer, an exchange of air masses takes place during late winter and in spring in the region of baroclinic zones and tropopause discontinuities of the mid-latitudes (*vid.* Newell, 1963). The residence time of the tritiated water in the stratosphere is of the order of years; inter-hemispheric mixing in the stratosphere seems to occur on a similar time scale. The residence time of water in the lower troposphere, on the other hand, is of the order of 5–20 days, this being a short period relative to the north-to-south mixing rates, but within the time scale of circum-polar motions. As a result Tritium is deposited onto the face of the earth (by rainout or exchange with open water surfaces), more or less within the latitudenal band of its penetration from aloft.

More details can be found in reviews by Begemann and Libby (1957), Begemann (1958), Bolin (1958), Libby (1961 and 1962), Eriksson (1962), Ehhalt (1971),Gat (1980) and Michel (2007), among others.

During the cosmic-ray period, prior to the nuclear testing era, where the average annual input into hydrologic systems could be assumed at a constant level, the dating of the residence times of the meteoric waters in these systems was a rather straightforward procedure, limited to ages commensurate with the decay rate of the Tritium, i.e. to some tens of years, and the hydrological structure of the respective water body. The simplest case is that of a water body isolated from interchange with the environment after deposition, such as a bottled sample of precipitation or an agricultural product. In this case, the ratio of the Tritium activity of the sample (N_t) to that of the input (N_0) is simply given by the radio-active decay law, (λ being the radioactive decay constant which is related to the radioactive half-live, $T^{1/2}$, by the relationship: $\lambda = \ln 2 / T^{1/2}$)

$$N_t/N_0 = e^{-\lambda i}$$

from which the "age" (t)of the sample is derived:

$$t = -\lambda \cdot \ln(N_t/N_0)$$

A similar equation would relate the inflow to outflow concentration of an ideal piston-flow system, such as a confined aquifer; in this case, the derived age is a measure of the through-flow time. In reality, due to partial mixing and hydrodynamic dispersion during the flow through the system, a correction has to be applied depending on the hydro-dynamic characteristics of the flow, as discussed by Nir (1964).

For any steady state system, one can define an age function, Φ_i, which details the fraction of water residing in the system for each of the i years counting backwards, so that $\Sigma\Phi_i = 1$. For an open, well-mixed through-flow system such as a shallow lake, the age function has the form of an exponentially decreasing one (the EMS model, *vid.* Zuber, 1986), so that $\Phi_i = e^{-ki}$ and k is inversely proportional to the mean residence time of the system.

It is to be noted, however, that over an open water surface, the Tritium input into the system is provided not only by the precipitation and river inflow but also by exchange between the surface waters and the overlying atmospheric moisture, so that the Tritium residence time is not strictly equal to that of the water. Pertinent studies of this are by Oestlund and Berry (1970) and Imboden *et al.* (1977) on Lake Tahoe, where it was found that 68% of the Tritium input to the lake was the result of molecular exchange with the ambient moisture while only 32% was added by precipitation.

Once the anthropogenic additions in the 1950-ties and 60-ties grossly perturbed the natural Tritium levels, the rules of the game changed completely since the basic assumption for dating the waters, i.e. a relatively constant input, was not realised anymore. However, the large and notable tracer peaks of the early 60's opened up possibilities of utilising this peak to follow its passage and dissemination through the hydrologic systems, especially large groundwater bodies (Eriksson, 1958).

With the declining atmospheric Tritium levels in recent years, the dating of water bodies by the Tritium method has become ambiguous. To illustrate this ambiguity, consider an area where at the moment of measurement (say in the year 2000), the level of Tritium in precipitation has been reduced to 10 TU; a value of close to 10 TU in a water sample can then be interpreted as either a direct recharge or that it represents the residue of about 5% of the input from the time of the peak concentrations in the 1960's.

An elegant way out of this quandary is provided by the measurement of both Tritium and its decay product ^{3}He (by advanced mass-spectrometry, as described by Schlosser *et al.* 1988 and Solomon *et al.* 1992). In a closed system, the latter integrates the total amount of Tritium that had decayed from the time of closure to the time of measurement. The age of the water in such a closed system is given by the equation:

$$t = \lambda \cdot \ln\{1 + [^3\text{He}]/[^3\text{H}]\}$$

where $[^3\text{He}]/[^3\text{H}]$ represents the atom ratio of Tritium and its decay product ^{3}He. The use of this methodology for open throughflow systems is discussed by Carmi and Gat (1992).

Athavale, R. N., Lal, D. and Rama, S. (1967). The measurement of Tritium activity in natural waters, II. Characteristics of global fallout of ^{3}H and ^{90}Sr. *Proc. Indian Acad. Sciences* 65: 73–103.

Begemann, F. and Libby, W. F. (1957). Continental water balance, groundwater inventory and storage times and worldwide water circulation patterns from cosmic rays and bomb Tritium. *Geochim. Cosmochim. Acta* 12: 277.

Begemann, F. (1958). New measurements on the worldwide distribution of natural and artificially produced Tritium. In: *Proc. 2nd Internat. Conf. on the Peaceful Uses of Atomic Energy* 18: 545.

Bigeleisen, J. (1962). Correlation of Tritium and Deuterium isotope effects. *Tritium in the Physical and Biological Cycles* IAEA, Vienna. 1: 161–168.

Bolin, B. (1958). On the use of Tritium as a tracer for water in nature. *Proc. 2nd Internat. Conf. on the Peaceful Uses of Atomic Energy* 18: 336.

Brown, R. M. and Grummit (1956). The determination of Tritium in natural waters. *Can. J.Chem.* 34: 220–226.

Buttlar, H. V. and Libby, W. F. (1955). Natural distribution of cosmic-ray produced Tritium, 2. *J. Inorg. Nuclear Chem.* 1: 75–95.

Carmi, I. and Gat, J. R. (2000). Estimating the turnover time of groundwater reservoirs by the Helium-3/Tritium method in the era of declining atmospheric Tritium levels; opportunities and limitations in the time bracket 1990–2000. *Israel J. Earth Sci.* 43: 249–253.

Eriksson, E. (1958). The possible use of Tritium for estimating groundwater storage, *Tellus* 10: 472–478.

Eriksson, E. (1962). An account of the major pulses of Tritium and their effect in the atmosphere. *Tellus* 17: 118–130.

Ehhalt, D. H. (1971). Vertical profiles and transport of HTO in the troposphere. *J. Geophys. Res.* 76: 7351–7367.

Gat, J. R. (1980). The isotopes of hydrogen and oxygen in precipitation. Chapter 2, *Handbook of Environmental Isotope Geochemistry* (Fritz and Fontes, eds.) Vol 1, Elsevier.

Grosse, A. V., Johnston, W. M., Wolfgang, R. L. and Libby, W. F. (1951). Tritium in nature. *Science* 113: 1–2.

Imboden, D. M., Weiss, R. F., Craig, H., Michel, R. L. and Goldman, Ch. R. (1977). Lake Tahoe Geochemical Study: 1. Lake chemistry and Tritium mixing study, *Limnology and Oceanography* 22: 1039–1051.

Kaufmann, S. and Libby, W. F. (1954). The natural distribution of Tritium. *Phys. Rev.* 93: 1337.

Lal, D. and Peters, B. (1967). Cosmic ray produced radioactivity on the earth. *Handbuch der Physik*, Springer: 46/2: 551–612.

Libby, W. F. (1953). The potential usefulness of natural Tritium. *Proc. Natnl. Acad. Sciences USA* 39: 245–247.

Libby, W. F. (1961). Tritium Geophysics. *J. Geophys. Res.* 66: 3767–3782.

Libby, W. F. (1962). Tritium Geophysics: recent data and results. In: *Tritium in the Physical and Biological Cycles* IAEA, Vienna. 1: 5–32.

Lucas, L. L. and Unterweger, M. P. (2000). Comprehensive review and critical evaluation of the half-life of Tritium. *J. Res. Natnl. Institute Stand. Technology.*

Michel, R. L. (2007). Tritium in the hydrologic cycle. *Isotopes in the Water Cycle* (edited by Aggarwal, Gat and Froehlich). Springer: pp. 53–66.

Newell, R. S. (1963). Transfer through the troposphere and within the stratosphere. *Quat. J. Royal Met. Soc.* 89: 167–204.

Nir, A. (1964). On the interpretation of Tritium 'age' measurements of ground-water. *J. Geophys. Res.* 69: 2589–2595.

Nir, A., Kruger, E. T., Lingenfelter, R. E. and Flamm, E. J. (1966). Natural Tritium. *Rev. Geophys.* 4: 441–456.

Oestlund, H. G. and Berry, E. X. (1970). Modification of atmospheric Tritium and water vapor by Lake Tahoe, *Tellus* 22: 463–465.

Schlosser, P., Stute, M., Dorr, H., Sonntag, Ch. and Munnich, K.O. (1988). Tritium/^{3}He dating of shallow groundwater. *Earth Planet. Sc. Lett.* 69: 353–362.

Solomon, D. K., Poreda, R. J., Schiff, S. L. and Cheng, J. A. (1992). Tritium and Helium-3 as groundwater age tracers in the Borden aquifer. *Water Resources Res.* 28: 741–755.

Zuber, A. (1986). Mathematical models for the interpretation of environmental radio-isotopes in groundwater systems. *Handbook of Environmental Isotope Geochemistry* (P. Fritz and J. Ch. Fontes, eds.), Elsevier, 2: 1–55.

Index